Operating Large Scale Membrane Bioreactors for Municipal Wastewater Treatment

Operating Large Scale Membrane Bioreactors for Municipal Wastewater Treatment

Christoph Brepols

IWA
Publishing
London • New York

Published by **IWA Publishing**
Alliance House
12 Caxton Street
London SW1H 0QS, UK
Telephone: +44 (0)20 7654 5500
Fax: +44 (0)20 7654 5555
Email: publications@iwap.co.uk
Web: www.iwapublishing.com

First published 2011

Cover design by www.designforpublishing.co.uk
Typeset by MPS Limited, a Macmillan Company, Chennai, India, www.macmillansolutions.com
Printed by Lightning Source

British Library Cataloguing in Publication Data
A CIP catalogue record for this book is available from the British Library

Library of Congress Cataloging-in-Publication Data
A catalog record for this book is available from the Library of Congress

ISBN13: 9781843393054
ISBN:1843393050

Contents

Contributors

Throughout the years, a number of organisations and individuals have contributed to the success of the three Erftverband MBRs.

We would like to thank the Environment ministry of the German federal state of North Rhine Westphalia for their financial support of the MBRs and the funding received in several research projects dedicated to the optimization of plant operations and the European Commission for funding research under the EUROMBRA project, which was a research project under the Sixth Framework Programme (Priority "Global Change and Ecosystems"), Contract No. 018480 – EUROMBRA.

Following organisations also contributed to research and development activities at the Erftverband plants. Aachener Verfahrenstechnik AVT and Institut für Siedlungswasserwirtschaft ISA at the RWTH Aachen University, Institut WAR at the TU Darmstadt, Institut für Siedlungswasserwirtschaft at the TU Braunschweig, IWW Rheinisch Westfälisches Institut für Wasser Beratungs- und Entwicklungsgesellschaft at the Duisburg-Essen University and Fachgebiet Siedlungswasserwirtschaft at the Kassel University.

We would also like to thank the plant operators at Rödingen, Nordkanal and Glessen for taking on the challenge of running a new technology.

The authors
Christoph Brepols, Kinga Drensla, Norbert Engelhardt, Torsten Engels, Andreas Janot, Heinrich Schäfer, Michael Trimborn

Erftverband
Am Erftverband 6
50126 Bergheim
Germany
info@erftverband.de

Chapter 1
Introduction

Christoph Brepols and Norbert Engelhardt

The protection of the raw material water is one of the main tasks of today and of tomorrow. Clean water as vital foodstuff for humanity and, as basis of an intact environment, is guaranteed not the least by a comprehensive treatment of wastewater. With this, the discharge of treated wastewater into rivers and streams from wastewater treatment plants is today often subject of strict quality requirements.

The Erftverband has committed itself to this task within the catchment area of the Erft river in Germany. As a regional water association under public law the Erftverband pursues innovative and integrative approaches towards water related issues in three key operational areas: the investigation of water management relationships for the protection of the water supply, the expansion and the care of flowing waters and the treatment of wastewater.

WASTEWATER TREATMENT WITH MEMBRANE BIOREACTORS

From a global perspective a growing confidence in MBR technology is demonstrated by the exponential increase in the cumulative MBR installed capacity. With new factors coming into play, MBR technology is now beginning to mature such that the market is expected to grow substantially over the next decade. Evidence suggests that MBRs will continue to penetrate further into the effluent treatment market, with the number of players in the global market increasing. Currently, the market is dominated by the two leading companies Zenon and Kubota. Whilst the domination of these two companies is likely to continue in the short to medium term, the global demand for the technology is such that a broader range of products is likely to be sustainable in the future, in particular if individual products are tailored towards niche market applications (Judd, 2006).

Recent reports from the Americas (Yang *et al.*, 2006; Mokbel *et al.*, 2009), South-Africa (Erwee *et al.*, 2009), Australia (Bagg, 2009) and Asia (Hashimoto *et al.*, 2009; Tao *et al.*, 2009) provide evidence of the ever-increasing interest in the technology and a growing number of applications worldwide.

In 2007 Lesjean *et al.* recorded 409 MBR references in Europe alone, among which approximately one quarter was related to municipal references (111 plants) and three quarters to industrial applications (298 units). The industrial market was the pioneer market in the early 1990's, whereas the municipal market really started to kick off in 1999 (Figure 1.1). Both municipal and industrial market witnessed a sharp increase, corresponding to the commercial success and broad applications of the immersed MBR technologies. It is expected that this progression rate will be at least sustained in the next years. A further acceleration of the rate will depend on improved competitiveness of the technology and the evolution and implementation of European and national regulations.

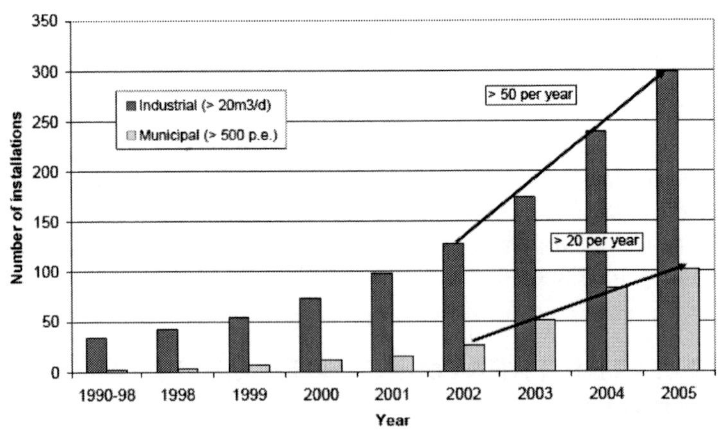

Figure 1.1 Development of the industrial and municipal MBR market in Europe (Lesjean *et al.*, 2007)

Package plant MBRs with sizes from 4 to 40 population equivalents (PE) have also been installed in larger numbers, many of them in environmentally sensitive areas. In some cases MBR is the only technology, that is approved by local environmental authorities for decentralized wastewater treatment (Busse, 2009).

Industrial MBR applications are well known for the treatment of landfill leachate and also, among others, for wastewater treatment in the pharmaceutical and food and beverage industry (Rosenwinkel *et al.*, 2000). Maritime use on board large ships is another field of application. The nature of the wastewater, the demand for additional treatment, water reuse, site restrictions and the opportunities to retrofit existing WWTPs can be seen as major drivers towards MBR technology in the industrial sector.

WATER MANAGEMENT IN THE OPERATIONAL AREA

The Erftverband is one of the special statutory water associations in North-Rhine Westphalia, Germany. In its Association area, it currently operates 44 wastewater treatment plants with capacities of from 500 to 132,000 PE. With this it is responsible for the treatment of the wastewater of more than one million individuals.

Like those in the association area, the great majority of German wastewater treatment plants have been upgraded in recent years in accordance with the demands of the Wastewater Ordinance (AbwV), and in particular have reached a high purification level with regard to the elimination of the nutrients nitrogen and phosphorous (Steinmetz H., 2008). The overall annual investment in wastewater treatment infrastructure has been relatively stable although it decreased from € 6.9 billion in 2000 to € 4.6 billion in 2006. This can be explained by the high investment that has already been completed to reach the goals of the EU legislation (ATT *et al.*, 2008).

Nearly 10,000 municipal WWTPs exist in Germany. A total of 96% of all inhabitants are connected to a public sewer system. From a total of 9.4 billion m^3 of domestic wastewater 97% are treated by biological nutrient removal (BMU, 2010a) which corresponds to the highest standard of the EU directive (91/271/EWG) on the treatment of municipal sewage. The remainder of the household wastewater is mostly treated in decentralised small size or package plant WWTPs. After the first full-scale municipal MBR installations went into operation, their number increased to at least 15 with capacities from 700 PE to 80,000 PE (Table 1.1). A great many of them are to be found in the federal state of North Rhine Westphalia (NRW). The water resources are under stress from competing industrial, agricultural and civic utilisations. Although water scarcity is uncommon, maintaining a high water quality is one of the major goals of the national environmental policy. In particular, MBRs have been established in specific hydrological situations where pathogens are of special concern because receiving waters are used for bathing and recreation or contribute directly to a drinking water reserve. Besides the hydrological situation, the legislative framework may be seen as a trigger for the application of MBR.

Table 1.1 MBRs for municipal wastewater treatment in Germany

Plant	Treatment capacity (PE)	Membrane supplier	Year of commissioning
Nordkanal	80000	GE Zenon	2004
Hutthurm	22000	Huber VRM	2008
Makranstädt	12000	GE Zenon	2000
Eitorf	11625	Kubota	2005
Seelscheid	10500	Kubota	2004
Monheim	9700	GE Zenon	2003
Konzen	9700	Kubota	2007
Glessen	9000	GE Zenon	2008
Rurberg	6200	Kubota	2005
Rödingen	3000	GE Zenon, KMS Puron	1999
Schramberg	2600	GE Zenon	2004
Xanten-Vynen	2000	A 3 GmbH	2005
Büchel	1000	Kubota	1999
Knautnaundorf	900	Martin systems AG	2001
Simmerath	700	KMS Puron	2003

The Erftverband introduced Germanys first MBR for municipal wastewater treatment in 1999. A few years later, in 2004, the Nordkanal plant was launched, which at that time was the world's largest municipal MBR. The third Erftverband MBR was completed in 2008.

The Erft river, after which the association is named, flows into the river Rhine northwest of Cologne, Germany. The Association area covers almost 2000 square kilometres. It lacks large industrial areas, but it is characterised by the extraction and conversion into energy of Rhenish brown coal. The open-cast lignite mining has a significant impact on the hydrology of the whole region. Groundwater levels are low and for many rivers the effluent of the WWTPs is often the only constant supply of water. At the same time the region is densely populated and intensively used for crop production. Biological nutrient removal has been generally established at the municipal WWTPs during the past 20 years. Additionally, one third of the WWTPs under the Erftverbands authority employ tertiary filtration to meet stringent discharge criteria. The increasing cost for high quality wastewater treatment and the limitations of conventional activated sludge technology spurred the initial interest in MBR technology.

Membrane technology has been publicly perceived as a future key technology of the water sector. Besides environmental policies on water quality, capacity building and the endorsement of new technologies in the water sector can be seen as main strategic objectives for the public funding of MBR technology received in Germany and in NRW in particular (Uhlenberg, 2007). Earmarked assets from the wastewater levy (AbwAG) are used to fund measures that improve water quality. With a focus on innovative technologies, MBR projects also received a share of the funding which initially reduced the end-users economic risk of investing in a new technology. As MBR technology emerged from its early stage, the amount of funding has been gradually reduced over the years.

Considering the high standard and degree of wastewater treatment, the domestic market for municipal wastewater treatment in Germany can be considered as relatively mature. Therefore, opportunities for the application of new MBRs in Germany, be it as greenfield or retrofitted installations, are not very widespread. Although MBRs have well-known advantages, they also must compete against other well-established treatment technologies such as, for example, UV disinfection or tertiary filtration. It is then often an inclination towards innovative technologies together with the strict requirements on water quality enforced on a regional level, that foster the application of the MBR technology.

THE ROLE OF END USERS IN THE MEMBRANE BUSINESS

Design, construction and operation of MBR facilities has not yet become a routine function. Although consultants, suppliers and construction companies offer their know-how to promote the proper application of MBR facilities world-wide, it should not be ignored that, in particular, the long-term success of the application lies finally within the responsibility of the end user as the awarding authority and often also the operator of the plants. A variety of factors must be taken into account.

(a) MBR technology offers a whole array of possible options for choosing a suitable plant layout. The size of the installation, effluent requirements and type of membrane are key factors.

(b) Construction details may differ significantly from conventional technologies and influence the proper operation of the membrane modules and the biological process as well.

(c) The start up phase requires an effort in the adjustment of all settings. Improper operation may hamper membrane integrity and performance straight away. The operational staff thus has to be trained to run the plant with prudence and precision.

(d) Membrane filter performance and energy consumption are critical in managing cost. There may be trade-offs between possible design options and cost effectiveness, but unjustified savings in construction, pretreatment or membrane cleaning may result in overly reduced membrane service life, higher energy consumption or higher labour costs.

(e) Worldwide research and development activities produce both incremental improvement and considerable change in MBR technology. It is essential to reflect on possible changes and arrange for improvements in all stages of project development.

(f) Once an MBR plant is in operation, it should typically run for years or even decades. As membranes are still the most expensive part of the equipment, a key to long-term economic success is the membrane life cycle. Service life can be enhanced through stability in operating conditions and plant performance, but then also the question of interchangeability of membrane modules arises.

The Erftverband has made great strides in meeting these challenges. It can be shown that the three MBR plants are a success. Although the Nordkanal plant is still among the world's largest MBR facilities, it is fully anticipated that, in the meantime, much larger ones have already been built or will be completed in the near future. The advent of such projects validates the worth and the prospect of success for this innovative technology.

REFERENCES

Arbeitsgemeinschaft Trinkwassertalsperren e. V. (ATT), Bundesverband der Energie- und Wasserwirtschaft e. V. (BDEW), Deutscher Bund der verbandlichen Wasserwirtschaft e. V. (DBVW), Deutsche Vereinigung des Gas- und Wasserfaches e. V., Technisch-wissenschaftlicher Verein (DVGW), Deutsche Vereinigung für Wasser wirtschaft, Abwasser und Abfall e. V. (DWA), Verband kommunaler Unternehmen e. V. (VKU): Branchenbild der deutschen Wasserwirtschaft 2008 *(Business profile of the German water industry)*, wvgw Wirtschafts- und Verlagsgesellschaft Gas und Wasser mbH, Bonn, Germany.

Bagg W. K. (2009). Infrastructure optimisation via MBR retrofit: a design guide. *Water Sci. Technol.*, **59**(2), 323–330.

BMU, Bundesministerium für Umwelt, Naturschutz und Reaktorsicherheit (2010): Gewässerschutz – Abwasser – Private Haushalte – Statistik *(Water protection – Wastewater – Private Households – Statistics)*, http://www.bmu.de/gewaesserschutz/fb/abwasser_priv_haushalte (accessed 19 January 2010).

Busse-IS GmbH (2010): Referenzanlagen BusseMF in Gewässerschutz- und Naturschutzgebieten *(Reference plants BusseMF in water and nature reserves)*, http://www.busse-is.de (accessed 19 January 2010).

Erwee H. and Gademan G. (2009). Sustainable Waste Water Treatment: Membrane Bio Reactor (MBR) Plant: Sedgefield, South Africa. *International Wisa Membrane Technology Conference 2009*, 13–15 May 2009, Stellenbosch, Western Cape, South Africa.

Hashimoto T., Suzumura K., Itokawa H. and Murakami T. (2009). Study of MBR system with separate membrane tank for the reconstruction of large-scale WWTPs. *5th IWA Specialised Membrane Technology Conference for Water and Wastewater Treatment*, 1–3 September 2009, Beijing, P.R. China.

Judd S. (2006): *The MBR Book, Principles and Application of Membrane Bioreactors in Water and Wastewater Treatment*, Elsevier, Oxford, UK.

Lesjean B. and Huisjes E. H. (2007). Survey of European MBR market, trends and perspectives. *IWA 4th International Membrane Technologies Conference*, 15–17 May 2007, Harrogate, UK.

Mokbel A. and Sen D. (2009): Design and Optimization of MBR Systems for Residential Wastewater Treatment and Reuse in Mexico and Emerging Markets using Aquifas, *WEFTEC conference proceedings*, 10–14 October 2009, Orlando, Florida, USA.

Rosenwinkel K.-H., Wagner J. and Nagy J. (2000). Membranverfahren in der industriellen Abwasserbehandlung. *Chemie Ingenieur Technik – CIT*, **72**(5), 433–440.

Steinmetz H. (2008). *Current Developments and Perspectives in Municipal Wastewater Disposal in Germany and Worldwide, Water and Waste*, Special edition, GWV Fachverlage GmbH, Wiesbaden, Germany.

Tao G., Kekre K., Htun Oo, M. Viswanath B., Lew C. H., Kan L. M. and Seah H. (2009). Large scale membrane bioreactor plant design (retrofit) and optimisation. *5th IWA Specialised Membrane Technology Conference for Water and Wastewater Treatment*, 1–3 September 2009, Beijing, P.R. China.

Uhlenberg E. (2007): Minister für Umwelt und Naturschutz, Landwirtschaft und Verbraucherschutz des Landes Nordrhein-Westfalen, „Für eine gesunde Umwelt – die Wasserwirtschaft in Nordrhein-Westfalen, ein Wirtschaftsfaktor und Wirtschaftsmotor.", Rede zum 1. Deutscher Wasser- und Energietag *(Minister of Environment and Nature Conservation, Agriculture and Consumer Protection of North Rhine-Westphalia, 'For a healthy environment – the water industry in North Rhine-Westphalia, an economic factor and an economic engine.' Speech on the first German Water and Energy Day)* 06.02.2007, Essen, Germany.

Yang W., Cicek N. and Ilg J. (2006). State-of-the-art of membrane bioreactors: worldwide research and commercial applications in North America. *J. of Membr. Sci.*, **270**(1–2), 15 February 2006, Pages 201–211.

Chapter 2

Design and process configuration of MBR plants

Christoph Brepols and Heinrich Schäfer

INTRODUCTION

Membrane bioreactor technology integrates membrane filtration into the activated sludge process for municipal wastewater treatment. The use of the filters which can be submersed inside the bioreactor or arranged in dedicated external vessels has extensive effects on the process and the design of WWTPs. Two core benefits result:

(1) High effluent quality: By using microfiltration membranes, all suspended solids can be retained from the outfall. Not only the organic and nutrient load can thus be reduced, but all germs and even viruses are held back. This process opens the way for water reuse and, without further treatment, the effluent for example satisfies the requirements of the EEC directive on bathing water quality.

(2) Low footprint: The microorganisms for biological wastewater treatment are retained completely within the bioreactor. This leads to increased concentrations of active biomass, so bioreactor size can be reduced. The filters outperform the usual settling tanks as phase separation units, so no secondary clarifiers and additional treatment steps are required. This leads to a very compact layout for the plants.

Nevertheless, the employment of membrane filters in municipal wastewater treatment faces several difficulties due to the characteristics of the wastewater and the process as a whole:

(1) high inflow with low waste content compared to industrial applications;
(2) high diurnal and seasonal variations of temperature, flow and loading;
(3) high amounts of fat, hair and fibrous material in the raw wastewater.

These factors need to be reflected in the design of MBR plants.

PRETREATMENT

Domestic sewage carries a high amount of fibrous and coarse waste. To protect membrane filters from the adverse effects of accumulation of unwanted materials, considerable effort has to be made to pretreat the sewage mechanically.

In the beginning, the importance of mechanical pretreatment was underestimated. Originally the Rödingen MBR employed only a twin channel step screen with 3 mm bar width. This very soon led to serious problems with filter blocking, clogging, increased fouling and filter damage by rough materials. The pretreatment had to be retrofitted.

The Erftverband today favours multi-staged concepts on the plants with additional fine screens or sieves removing the undesirable material, as shown in Table 2.1. The screens and sieves are designed with reserve machinery capacity to achieve a fail-safe operation of the treatment. The full hydraulic capacity can still be achieved with one machine out of operation. Therefore, either twin machinery (Figure 2.1) or emergency overflows, that are connected to a buffer tank have been designed. In any case precautions are taken to prevent raw wastewater from passing the pretreatment step without at least partial treatment.

Table 2.1 Plant specifications, mechanical pretreatment

Rödingen, 3,000PE	Glessen, 9,000PE	Nordkanal, 80,000PE
Bar screen 3mm; twin channel	Bar screen 6mm; single	Bar screen 5mm; twin channel
	20mm emergency overflow screen	
Aerated sand/grit chamber, One macerator	Aerated sand/grit chamber	Aerated sand/grit chamber
Bypass sieve 0.5mm for sludge screening	Mesh sieve 1.5mm, twin channel	Mesh sieve 1.5mm, twin channel, 1.0mm grid sieve (emergency overflow)

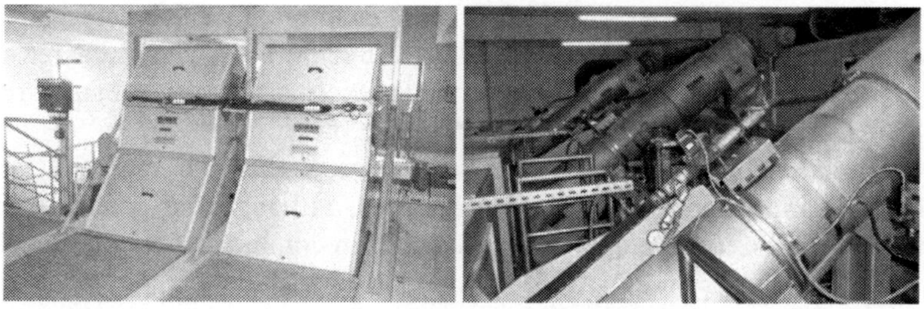

Figure 2.1 Nordkanal MBR, twin channel bar screen (left) and twin mesh sieves (right) with by-pass emergency sieve in the background

The recycle water from the screenings washing or the screenings press is fed into the influent of the pretreatment stage again, to avoid an eventual bypass of debris.

BIOREACTOR AND PROCESS LAYOUT

The biological treatment process at the Erftverband MBRs was designed as simultaneous aerobic sludge stabilization with low F:M ration and sludge retention times (SRT) of more than 25 days. This is also common for many conventional activated sludge plant (CAS), but MBR then differ widely in terms of sludge concentration and hydraulic retention time (Table 2.2).

Table 2.2 Quantities of key design parameters

Parameter	Rödingen	Glessen	Nordkanal	CAS
Sludge concentration, kg MLSS/m^3	12–18	8–12	12	3–5
F:M ratio, kgBOD/(kgMLSS d)	0.04	0.03	$<$0.05	$<$0.05
Hydraulic retention time (HRT), h	3.6	6.2	4.6	$>$14
Sludge retention time (SRT), d	25	$>$25	25	$>$25

Since biomass concentration MLSS in MBRs is many times greater than in conventional reactors, the bioreactor volume can be dramatically reduced. However practical operation has shown that there is an upper limit, that cannot be exceeded without penalty. The initial experience at the Rödingen MBR has shown that, for the given membrane filters, sustained operation with MLSS concentrations of more than 15kg/m^3 lead to an increasing accumulation of stagnant biomass in the filter cassettes and thus to a loss in membrane performance. The companies' other MBRs were therefore designed with a 12kg/m^3 threshold in mind, which allows enough room for inevitable MLSS variations during actual operation. Nonetheless, the MBRs are sized about one-third of comparable conventional reactors. The hydraulic retention time (HRT) is thus reduced by 50–70%. This is the reason for a more pronounced dynamic performance of the biological treatment and it renders the MBR more susceptible to variations in the inflow.

The practical experience at Nordkanal shows that the required effluent quality is maintained under all circumstances, although the actual hydraulic residence time is considerably lower than the recommendation of Pinnekamp *et al.* (2006). Furthermore, the operation of the Rödingen plant has shown that even stricter effluent requirements can be also met at short minimum HRTs (Erftverband, 2004). The operational results at Glessen also indicate that full nitrification is achieved under peak conditions. Process configurations and effluent requirements are displayed in Table 2.3. The effluent concentrations have to be maintained as maximum peak concentrations also at low wastewater temperatures.

Table 2.3 Process configuration of the three MBRs

	Rödingen	**Glessen**	**Nordkanal**
Bioreactor volume	$400m^3$ + $80m^3$ (filtration tanks)	$1600m^3$ + $320m^3$ (filtration tanks)	$9300m^3$
Specific volume	160L/PE	213L/PE	116L/PE
Process configuration	Simultaneous aerobic sludge stabilization,	Simultaneous aerobic sludge stabilization	Simultaneous aerobic sludge stabilization,
Nitrogen removal	Intermittent or upstream denitrification	Intermittent denitrification	Upstream denitrification
Phosphorous removal	Simultaneous precipitation	Simultaneous precipitation	Simultaneous precipitation
Temperature	5°C	8°C	10°C
Effluent requirements	COD <30mg/L NH_4-N <4mg/L P_{tot} <1.5mg/L	COD <30mg/L NH_4-N <1.5mg/L P_{tot} <0.6mg/L	COD <90mg/L NH_4-N <10mg/L P_{tot} <1.5mg/L

MEMBRANE FILTRATION COMPARTMENT

The filtration units at the Erftverband are equipped with submersed hollow fibre (HF) microfiltration membranes. The filtration compartment of the large installation at Nordkanal MBR can be seen in Figure 2.2.

Figure 2.2 Filtration/nitrification compartment, Nordkanal MBR

There are two general ways of integrating submersed membranes into the activated sludge process. While the plants in Rödingen and Glessen have separate membrane filtration tanks the membrane filters in Nordkanal are inserted directly into the nitrification tank (Figures 2.3, 2.4 and 2.5).

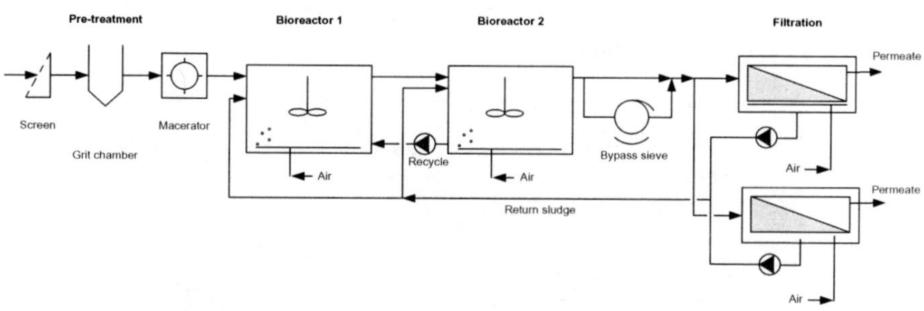

Figure 2.3 Process flow diagram Rödingen MBR plant

The filters are discontinuously operated, the hydraulic capacity of the filtration is adjusted to inflow variations partly by variations in the actual flux and to a large extent by shutting down single filtration lines temporarily at times of low inflow. The full hydraulic capacity is only required at maximum storm weather inflow. The Nordkanal and Glessen plant have been designed with redundant membrane area, such that the full hydraulic capacity can also be achieved when one or two filtration lines are shut down for cleaning or maintenance, by temporarily increasing the operating net flux of the remaining lines. Table 2.4 shows key data of the membrane filtration layout.

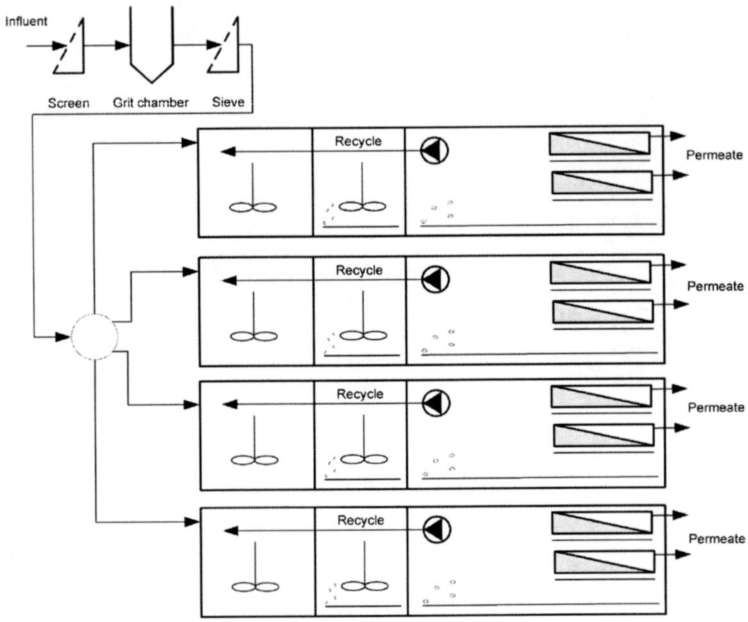

Figure 2.4 Process flow diagram, Nordkanal MBR

Figure 2.5 Aerial view of the retrofitted Glessen WWTP

Table 2.4 Membrane filtration layout

	Rödingen	Glessen	Nordkanal
Membrane area	5,280m^2	12,100m^2	84,480m^2
Membrane type	HF modules, ZeeWeed 500C/KMS Puron	HF modules, ZeeWeed 500D	HF modules, ZeeWeed 500C
Configuration	2 separate filtration tanks	4 separate filtration tanks	8 integrated filtration lines
Design Net flux	26L/(m^{2}h)	24–32L/(m^{2}h)	24–32L/(m^{2}h)
Blower capacity per filtration line	1150m^3/h	1000m^3/h	4250m^3/h

The operational experience at the three MBRs has finally shown, that three main objectives should be pursued when designing the filtration tanks:

(1) Occasional optical inspections of the submerged filtration units are inevitable. This can be either facilitated by having a crane to remove the filtration units from the basin or a basin that can be easily emptied and accessed by an operator.
(2) Inflow and outflow design of the membrane compartments should facilitate the complete distribution of the biomass along the filters, avoid stagnant zones in the reactor and avoid local MLSS gradients on the membrane filters.
(3) Often the geometry of the tanks can only harbour a single type of commercially available filtration modules. This reduces the flexibility in later membrane replacement. It can be recommended that the construction allows easy adjustments to suit other products at least of the hollow fibre or flat-sheet type.

DESIGN OF THE RÖDINGEN MBR

The first MBR plant was established in Rödingen, North Rhine Westphalia (NRW), in 1999 to satisfy new, stricter discharge criteria. Otherwise a major expansion of the existing conventional facility would have been necessary. After considering an array of options the Erftverband saw a chance that MBR technology could eventually meet the requirements at an acceptable cost. A pilot MBR plant was installed for a 5 month period. The results gave Erftverband and the regulating government authorities confidence to invest into what then became the first full-scale municipal WWTP in Germany using immersed membrane filtration.

The plant was designed as a large scale pilot to explore the prospects of membrane bioreactor technology and optimize the biological and filtration performance of the plant. Two equally equipped bioreactor tanks were built to allow a maximum flexibility in operating the biological process. Separate filtration tanks were intended to allow the optimization of the filtration and the biological step separately from each other. Various options for influent and recycle distribution within the process were made possible. At that stage energy efficiency was considered to be only of secondary importance. The plant was built at the site of the old facility. The disused old bioreactor was converted into a storage tank for raw wastewater or biomass that is occasionally used during maintenance and repair works on the new bioreactors or the filtration lines.

Catchment area	
Capacity	3,000PE
Connected area	22ha paved surface
Sewer system	combined system
Wastewater inflow	
Annual quantity of wastewater	150,000m^3
Maximum inflow, dry weather	56m^3/h
Maximum inflow, storm weather	135m^3/h
Designed influent loading	
Chemical oxygen demand COD	420kg/d
Nitrogen load TKN	40kg/d
Solid matter load SS_0	250kg/d
Phosphorus load P_{total}	10kg/d

The original plant layout was subjected to several modifications and optimizations during recent years. The process flow diagram of the current plant configuration can be seen in Figure 2.3 and an aerial view in Figure 2.6. The result of the pioneering effort is that today the Erftverband has been operating this plant for over 11 years. The equipment of the process units is listed in Table 2.5.

DESIGN OF THE NORDKANAL MBR

The new Nordkanal WWTP is located on the banks of the name-giving Nordkanal west of the original wastewater treatment plant. The Nordkanal is a disused artificial waterway that was intended to connect the rivers Rhine and Maas but was never completed. Construction works were already given up in the early 19th century.

The old WWTP located 2.5km away from today's plant location was dismantled and a new pumping station was built on that site. A 20mm coarse screen exist upstream of the pumping station to protect it from inflowing large debris. A disused rain water storage tank of the old facility was converted into an emergency buffer tank for the MBR. The original design for the new Nordkanal plant envisioned a conventional system with primary clarifiers, aerobic and anoxic bioreactors, secondary clarifiers, filtration, and a separate anaerobic sludge digestion. The experience with the operation of the Rödingen MBR plant, convinced the Erftverband to change the original concept and to design an MBR. The MBR plant occupies only about 50% of the land that a conventional plant would require. The smaller footprint saved about 20,000m^2 of ground. This accounted for saved investment costs for the land acquisition and also for reduced environmental compensations required by German legislation.

Figure 2.6 Aerial view of Rödingen MBR, two new bioreactors and shelter for membrane filtration and mechanical pretreatment (left) and remains of the old CAS (right)

The plant was tendered as a turn-key installation. After the tendering process the final choice was made for an MBR system with hollow fibre membranes installed directly in the main bioreactors and with separate tanks for recovery cleaning of the membranes. The Nordkanal MBR design is thus different from the Rödingen MBR, where the membranes are installed in separate filtration tanks. Erftverband's decision for the final Nordkanal design was, among others, based on a numerical flow simulation, verifying the full mixing in the bioreactor tanks without any relevant gradient of MLSS concentration over the length of the bioreactor.

The wastewater is treated mechanically, chemically and biologically in several steps. The new plant has the following treatment stages: fine screen, aerated grit chamber and grease trap, sieve, simultaneous precipitation and membrane bioreactor. The latter consists of three staged tanks for upstream denitrification, the facultative aeration and for nitrification. The membrane filtration units are integrated into the nitrification tanks (Figure 2.7). Every tank contains two independent filtration lines. The 24 cassettes of each line are arranged in three blocks installed on the side walls of the bioreactor. Additional mixers underneath and between these blocks facilitate MLSS distribution inside the vessels. The blowers for the membrane and bioreactor aeration are accommodated in a building next to the tanks. The surplus sludge yielded in the bioreactors is directly extracted from the nitrification basins and is stored temporarily in stacking tanks and dewatered using a centrifuge. The operations building contains the central control room, a laboratory for on-site wastewater analyses and a workshop.

Table 2.5 Main buildings and process units, Rödingen MBR

Process unit	Equipment	Capacity
Influent pumping station	two pumps	capacity 135m^3/h
Fine screen	two channel without bypass	screen width 3mm
Grit- and fat chamber	one channel	volume 14m^3
Sieve, bypass sludge recycle	one channel	mesh diameter 0.75mm
Bioreactor feed pumps	two pumps	capacity 135m^3/h
Bioreactor	two tanks	volume 800m^3
Membrane filtration	two channels, for 6 casettes each	tank volume 160m^3
Membrane cleaning	mobile separate intensive cleaning unit	volume 5m^3
Phosphorus precipitation	simultaneous	
Return sludge pumps	two pumps	capacity 400m^3/h
Buffer tank		volume 400m^3
Sludge storage	one gravity thickener	volume 400m^3
Sludge dewatering	off-site	

Catchment area	
Connection capacity	80,000PE
Connected area	1,235ha paved surface
Sewer system	Combined system

Wastewater inflow	
Annual quantity of wastewater	5,500,000m^3
Maximum inflow, dry weather	1,024m^3/h
Maximum inflow, storm weather	1,880m^3/h
Daily design loading	
Chemical oxygen demand COD	9,600kg/d
Biochemical oxygen demand BOD_5	5,250kg/d
Nitrogen load TKN	897kg/d
Solid matter load AFS	5,600kg/d
Phosphorus load P_{total}	123kg/d

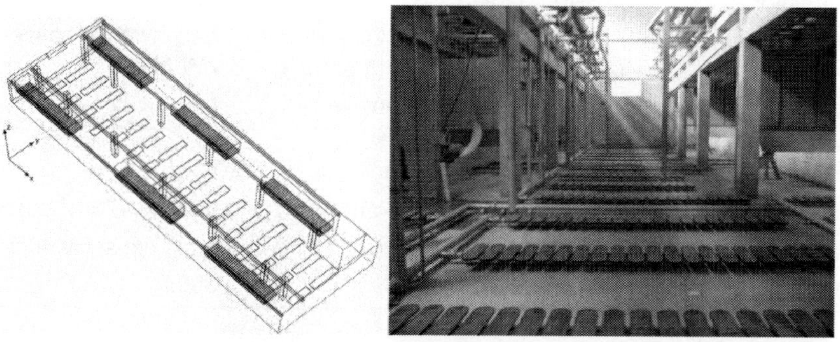

Figure 2.7 Schematic representation of a filtration/nitrification basin (left), (Hunze, 2005) and interior of an empty basin during construction (right) at Nordkanal MBR

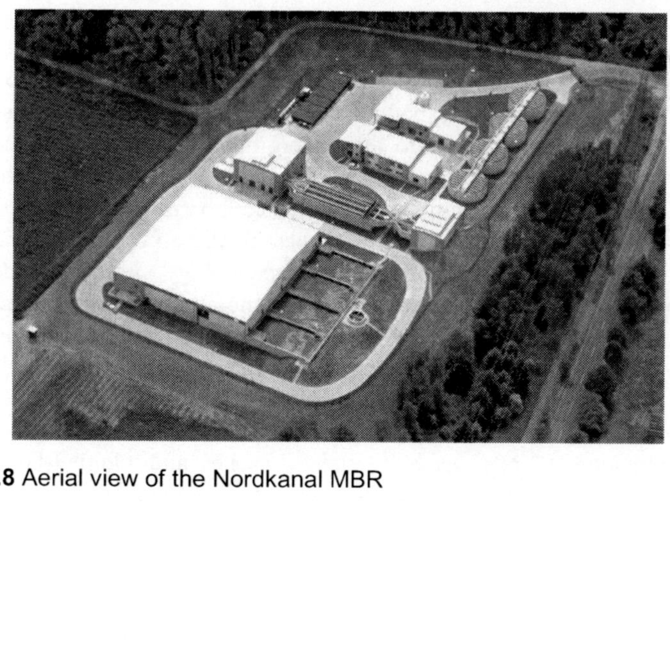

Figure 2.8 Aerial view of the Nordkanal MBR

Table 2.6 Nordkanal MBR, main buildings and process units

Process unit	Equipment	Capacity
Pumping station A	four pumps	capacity 1,600m^3/h
Pumping station B	three pumps	capacity 300m^3/h
Buffer tank		volume 2,000m^3
Pressure pipeline	two pipelines	length 2.5km
Fine screen	twin channel with bypass	screen width 5mm
Grit- and fat chamber	twin channel	volume 53m^3
Sieve	twin channel with bypass sieve	mesh diameter 1.5mm
Bioreactor	four tanks	volume 9,300m^3
Membrane filtration	eight channels	each with 24 cassettes ZW 500 C
Membrane cleaning	two separate intensive cleaning units	
Phosphorus precipitation	simultaneous	
Sludge storage	three gravity thickeners	volume 475m^3 each
Sludge dewatering	centrifuge	max. capacitiy 40m^3/h
Sludge liquor storage	one tank	300m^3

DESIGN OF THE GLESSEN MBR

The Glessen MBR is situated on the site of an old wastewater treatment plant next to the village of Bergheim – Glessen. The purified water is discharged into a stream, the Pulheimer Bach which, after a few kilometres, flows into the wetlands of the Pulheimer Laache. These wetlands are formed by a pervious gravel formation of the ancient river bed of the river Rhine. The wetlands have no effluent. The water percolates into the underground and thus contributes to the groundwater stream that is extracted at a nearby water works of the city of Cologne. As a means of precautionary resource conservation in the catchment area of the water works, an ultrafiltration MBR was finally accepted by the regulating environmental agency as the best available technology aiming at reducing the number of pathogens in the plant effluent and generally improving the river water quality. Another part of the effluent from Glessen is pumped back to the site of a disused WWTP at Fliesteden and feeds another stream, thus supplying the surface waters in the nearby nature conservation area Ommelstal. Improving the water quality of the two streams is an important measure to protect the regionally significant biotopes 'Große Laache', 'Kleine Laache', and 'Ommelstal' with their functions of nature conservation and local recreation (Figure 2.9).

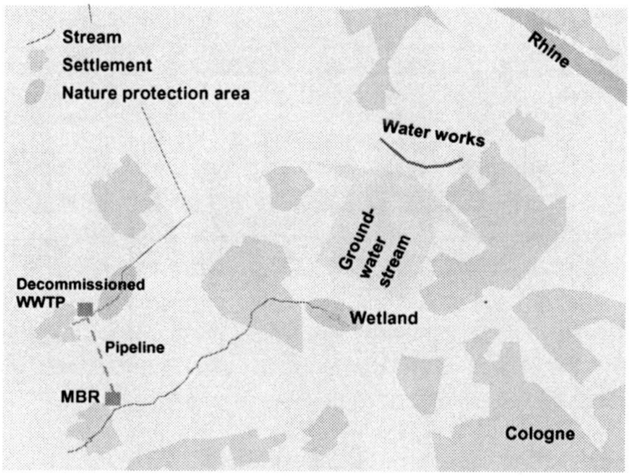

Figure 2.9 Hydrological situation of the Glessen MBR

The design concept of the new plant reflects the experience from the former two plants and was determined by strict quality and economical requirements. In retrofitting the existing CAS, the attempt was made to tap the full potential of membrane technology while integrating the existing structural fabric into the new plant. The wastewater, which was previously treated in two separate CAS plants, is today treated exclusively on the site of Glessen. MBR technology made it possible to retain the aeration tank of the old wastewater treatment plant and, at the same time, nearly double the capacity of the plant. This saved costs while improving the water quality significantly. Figure 2.5 shows an aerial view of the WWTP.

Because of the over-aged and insufficient equipment, the existing pretreatment facility had to be rebuilt entirely. The mechanical equipment of the bioreactor such as the surface aerators and agitators was replaced. A completely new automatic process control was applied. To adjust the treatment process to seasonal variations a flexible process control concept is used. The Glessen plant can be operated with either intermittent or simultaneous denitrification. Phosphorous removal is enhanced by chemical precipitation.

The membrane filters are installed in four newly built separate filtration chambers (Figure 2.10). Permeate pumps, cleaning equipment and blowers are installed in the adjacent building. The old operations building was renovated and slightly enlarged. The old secondary clarifier nowadays is used as a buffer tank. In case of extraordinary incidents, pretreated influent or a part of the biomass can be stored here. The sludge storage tank was left unchanged.

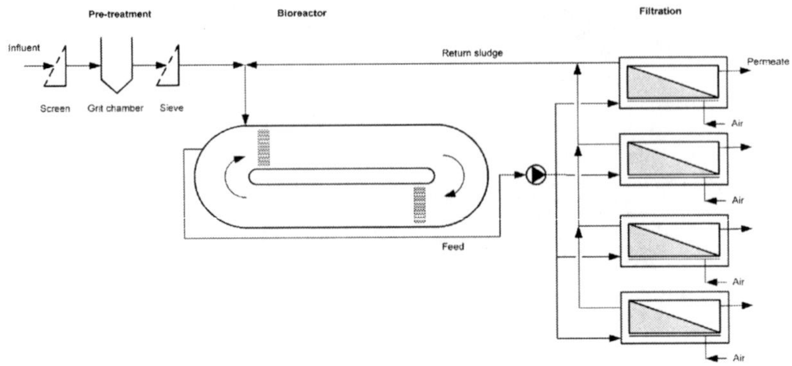

Figure 2.10 Process flow diagram, Glessen MBR

For the treatment of excess stormwater, which cannot be dealt with in the wastewater treatment plant following heavy rainfall events, the Erftverband has installed soil filter basins in the Glessen and Fliesteden drainage systems. Overflows from the sewer network are treated in artificial reed lagoons. The Erftverband has installed two new pressure pipelines for the treatment of the wastewater from the village of Fliesteden. The wastewater is pumped in one pipe from Fliesteden, where the old operations building has been converted into a pumping station, to the Glessen plant where it is treated and then pumped in the other pipe back to the stream near Fliesteden. The pipelines run over 3.6 kilometres under field tracks and farm land. Trenchless construction was used in order to protect nature and the landscape. Following dismantling of the plant at Fliesteden the site was reforested with broad-leaved trees typical for the location.

Catchment area	
Connection capacity	9,000PE
Connected area	52ha of paved surface
Sewer system	combined system
Wastewater Inflow	
Annual quantity of wastewater	900,000m^3
Maximum inflow, dry weather Q_d	156m^3/h
Maximum inflow, storm weather Q_r	268m^3/h
Daily design loading	
Chemical oxygen demand COD	1,008kg/d
Nitrogen load TKN	99kg/d
Phosphorus P_{tot}	15kg/d
Suspended solids SS_o	603kg/d

Table 2.7 Glessen MBR, main Buildings and process units

Process unit	Equipment	Capacity
Pumping station Glessen	three pumps	capacity 220m^3/h
Pumping station Fliesteden	two pumps	capacity 64m^3/h
Pressure pipeline	two pipelines	length 3.6km
Fine screen	one channel with bypass	screen width 6mm
Grit- and fat chamber	one channel	volume 53m^3
Sieve	twin channel	mesh diameter 0.75mm
Bioreactor	one tank	volume 1,400m^3
Membrane feed pumps	four pumps	capacity 1,000m^3/h
Membrane filtration	four channels, each with two cassettes ZW 500 D	tank volume 320m^3
Membrane cleaning	separate intensive cleaning unit	volume 23m^3
Phosphorus precipitation	simultaneous	
Buffer tank	one tank	volume 560m^3
Clean water pumping station	one pump	capacity 64m^3/h
Sludge storage	one gravity thickener	volume 500m^3
Sludge dewatering	off-site	

GENERAL DESIGN RECOMMENDATIONS

Nutrient Removal

If a MBR plant has to achieve full nitrification and at least partial denitrification, the process configuration for the nitrogen removal becomes decisive for the general layout of the biological reactors. The membrane filtration then can be combined with the biological treatment in several ways. These are different in their effect on the mass balances of the plants and thus need to be reflected when dimensioning the reactor compartments.

The membrane filtration may be directly integrated into the bioreactor e.g. the nitrification compartment or may be installed in separate filtration tanks outside the bioreactor.

For the submersed membrane filtration units, employed at the Erftverband's MBR plants the combinations of process configurations shown in Table 2.8 have been tested. The air-scouring of the membrane filters limits the options for the integrated filtration: The resulting oxygen concentration in the membrane compartment and the possible carryover to the anoxic zones limits the denitrification capacity. On the other hand, unintentional intermittent denitrification may occur in the filtration/nitrification tank while the filtration is shut down temporarily. All mass balances should be properly calculated when designing the plant, because of the effects of sludge recycling within the process.

Table 2.8 Combination matrix of denitrification processes with membrane filtration

MBR configuration	integrated filtration	separate filtration
Denitrification process	upstream	upstream simultaneous intermittent/alternating

Mass balances for different process configurations

The maximum allowable MLSS concentration at the filtration units limits the MLSS concentration in the bioreactor. The sludge recycle rate determines the MLSS concentration in the several compartments and thus the overall reactor volume of the installation.

As an example Figure 2.11 shows schematic representations of different reactor configurations – internal filtration with upstream denitrification (top), external filtration with upstream denitrification (middle) and external denitrification with simultaneous/intermittent denitrification (bottom) – and mass balances of the MLSS concentration in the reactor compartments.

The consequences on the mass balances of the plants can be illustrated by the following example: The maximum MLSS concentration at the membrane filtration units is 14g/L. The return sludge rate R_S is 400% and the recycle rate R_R is 280%. The resulting MLSS concentrations in the different reactor compartments are given in Table 2.9.

When dimensioning the denitrification compartment the oxygen carryover from the filtration compartments should be taken into account. Pinnekamp *et al.* recommend that the size of the denitrification compartment should be not less than 50% of the overall bioreactor volume to compensate the eventual losses of the actual denitrification capacity. As a result, the required bioreactor volume of an installation with external filtration tanks with upstream denitrification can be approximately 30 to 40% higher than for a plant with internal filtration.

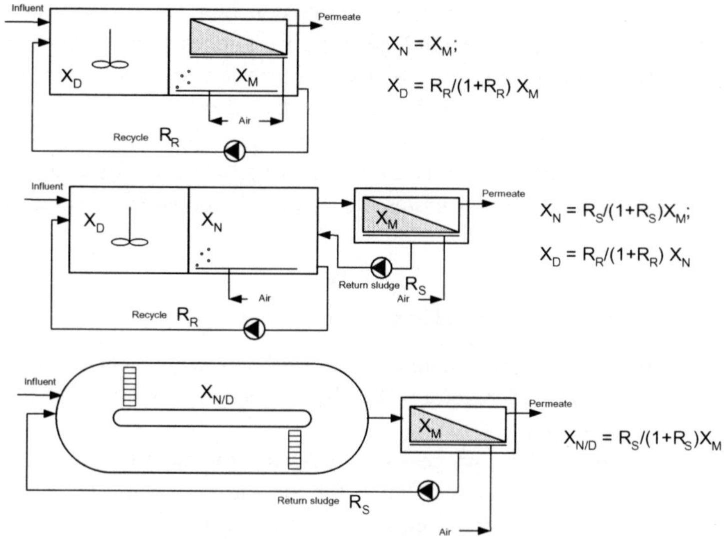

Figure 2.11 Schematic process flow schemes of MBR reactors

Table 2.9 Examples of MLSS concentrations for different reactor configurations

System		MLSS concentration, g/L	
Filtration	**Denitrification**	**Nitrification X_N**	**Denitrification X_D**
internal	simultaneous	11.2	11.2
internal	upstream	14	10.3
external	upstream	11.2	8.2

Inside or Outside Configuration

The advantages of an inside or outside configuration of the membrane filtration units have been discussed by De Wever *et al.* (2009). Decisions on the process configuration can be made on the basis of the following three aspects:

1. Size and footprint requirements of the plant. The inside configuration allows for smaller bioreactor volumes because of resulting higher MLSS in the bioreactors. The footprint can be reduced further because no additional filtration tanks have to be built.
2. Energy demand. With an inside configuration at least the energy for sludge recycling can be reduced. The cross flow aeration of the filters can be used for the aeration of the bioreactor as well. Especially for small MBRs there is often no mandatory additional aeration for the biological step.

3. Nitrogen removal. With the outside configuration return sludge and nitrate recycle can be held separately. The oxygen carryover to the denitrification zone is thus reduced. Filtration and biological treatment can be optimized separately. The increased overall volume makes the process less vulnerable to shock loadings.

Close interactions between pretreatment, biological and filtration processes open up a field of actuating variables that should be considered carefully when designing an MBR plant.

Projected Design Workflow

Based on the mass balance calculations the following work flow for the dimensioning of the bioreactor of MBR plants was developed at the Erftverband (Brepols *et al.*, 2005)

1. Determination of the process configuration for the nitrogen removal.
2. Calculation of the mass balance for the nitrogen removal, the required recycle rate and the fractions of anoxic and aerobic activated sludge.
3. Determination of the designed sludge retention time SRT, the surplus sludge extraction and the resulting total MLSS.
4. Determination of the maximum allowable MLSS concentration in the filtration compartment and of the return sludge rate.
5. Calculation of the MLSS mass balances for the different reactor compartments.
6. Dimensioning of the respective tank volumes.
7. Verification of the actual process conditions taking into account the effective tank volumes, aeration capacity, alkalinity etc.
8. Dimensioning of the membrane filtration including the peripheral equipment.

When dimensioning an MBR plant it is necessary to determine the required total amount of biomass necessary for the biological wastewater treatment and the allowable solids concentration on the membranes. For a sustainable long term operation 12 to maximum 15g MLSS/L seem to be appropriate. Influent loading, treatment requirements and biological process design are taken into account on the basis of common design principles with respect to an additional oxygen carry-over from the filtration units and the influence of the MLSS concentration on the aeration efficiency. When retrofitting existing WWTPs using membrane bioreactor technology the actual MLSS concentrations, the recycle and return sludge rates can be optimised e.g. reduced to make full use of existing reactor volumes.

SIMULATION AS A DESIGN TOOL

At different stages of the project development computational models have been used as tools for optimising the plant layout or the process operation. Biokinetic models were used to reveal insight into the process dynamics while, in the case of the Nordkanal MBR computational fluid dynamic models were also used to reveal and optimize the hydrodynamics of the system.

CFD Modelling

Many MBR plants employ filtration tanks that are strictly separated from the biological reactor. At Nordkanal the filtration units are directly inserted into the plants nitrification basins. This design was projected to reduce the footprint of the installation and consequently save cost for civil construction. However it raised several questions about the plants performance at the same time, as a reactor of this type and size had never been built before: (a) what are the flow characteristics of the reactor, (b) are there significant concentration gradients in the reactor, (c) are there hydraulic short circuits that reduce the biological treatment efficiency? To reveal insights into these topics, the engineering of the Nordkanal MBR was supported by CFD modelling.

The hydraulic residence time distribution of the nitrification/filtration tank was calculated by a stationary three dimensional model using the finite element method (Holthausen, 2003). Figure 2.12 shows the normalised result of a variant at full aeration and maximum inflow. The vertical line indicates the theoretical residence time as a quotient of reactor volume and inflow. The relative concentration of an alleged tracer substance indicates the residence time distribution. It could be seen that the HRT distribution approximately resembles a totally mixed reactor.

Another CFD simulation by Hunze (2005) validated that only a slight gradient in the MLSS concentration in the reactor could be expected while tracer simulation showed a tendency towards hydraulic short circuiting with the recirculation to the denitrification tank.

Tracer test studies at Rödingen MBR (Erftverband, 2001) have shown that the residence time distribution of an MBR with separate filtration tanks can also be similar to a totally mixed reactor, because of the high sludge circulation rates between the filtration units and the bioreactor. Brannock *et al.* (2007a, 2007b) came to similar conclusions for their reactor set-ups. A reason lies in the abundant mixing energy that is a result of the membrane air scouring.

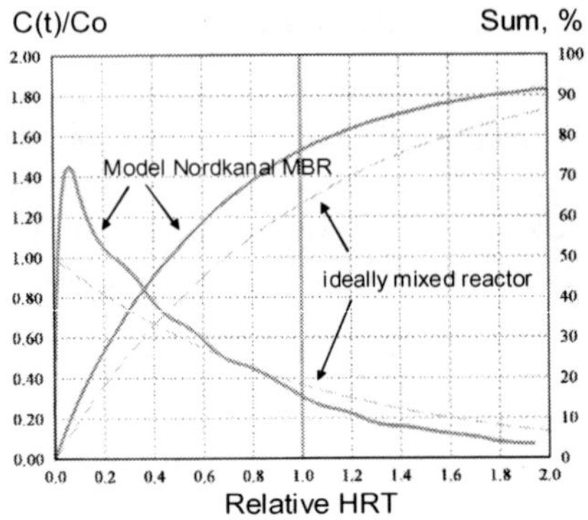

Figure 2.12 Hydraulic residence time distribution (Holthausen, 2003)

Biokinetic Modelling

Fenu *et al.* (2010) have synthesized previous studies on the use of activated sludge models (ASM) in MBR simulation and differentiated between unmodified and modified applications of ASM to MBR. Particular emphasis is placed on influent fractionation, biokinetics, and soluble microbial products (SMPs)/ exo-polymeric substances (EPS) modelling, and suggestions are put forward as to good modelling practice. For the unmodified ASM models, that also have been used by Erftverband, the authors come to the conclusion that these are of relevance in modelling MBRs as long as effluent characterisation, oxygen demand and sludge production are the concern. However, the comparison of biokinetic parameters from various studies to ASM default parameter values also revealed that the specific conditions present in MBRs may eventually lead to significant discrepancies.

Biokinetic models of the ASM family have been applied at the three Erftverband MBRs. The modelling aimed mainly at simulating the qualitative effects of plant operation and optimisation on the effluent concentrations. At Rödingen and Nordkanal, the models were used to investigate the observed plant behaviour after commissioning. At Glessen the model was also used as a tool to develop the process control strategies for the biological treatment during an earlier phase of the project. The simulations were carried out using the SIMBA software (IFAK) under MATLAB/SIMULINK. Throughout the years different versions of the software were in use.

Prior to the simulation of the Rödingen plant a simulation study of the pilot plant had been performed giving the required confidence that the strategies for model build-up and calibration known from various conventional activated sludge plant could be applied. The later model of the large scale plant was calibrated using the ASM 1 for several periods of up to a two month series and for operation with pre-denitrification and intermittent denitrification.

The Nordkanal model employs ASM 3 and was calibrated using on-line measurement data from the plant for flow, SAC-UV and Ammonia in the influent and effluent and Nitrate in the effluent.

For the Glessen MBR two model versions exist, using ASM2d and ASM3. While the models were build, actual performance data of the later MBR did not exist, so the calibration was based on the operational experiences of the existing CAS and information from former MBR simulations.

The Erftverband has developed a routine for building and calibrating CAS models that was also – with some modifications due to the intensity of the work – adopted for the MBR modelling. (1) Based on the geometry of the plant a hydraulic model is set up to resemble the flow characteristics. The control strategies for aeration, sludge recycle and sludge withdrawal are programmed. This set-up might be subject to modifications during the later calibration. (2) Based on existing measurement data, mass balances for the influent and effluent values are calculated. To increase the data-base, tailored measurement campaigns are carried out if necessary. (3) The influent COD characterisation is calibrated by trial-and-error. The biokinetic parameters are used according to Bornemann *et al.* (1996) and might get slightly modified in the course of a sensitivity analyses.

Flow Characteristics

The Rödingen model consist of two completely mixed reactor-blocks resembling the two bioreactors and a variable volume. In the original plant design, variations in the water level of the actual basins made up for up to 50% of the bioreactor volume. Two reactors with phase separation and biological activity resemble the separate membrane filtration compartments.

The Nordkanal MBR was reproduced by a hydraulic model consisting of a cascade of three totally mixed reactors (one virtual vessel for every existing bioreactor compartment with pre-denitrification, alternating zone, nitrification) with surplus sludge withdrawal and a phase separation with complete retention of all suspended model components and a sludge recycle.

The oxidation ditch of the Glessen MBR was represented by a cascade of 8 completely stirred reactors with a high internal recycle rate so that the gradient in oxygen concentration can be modelled (Figure 2.13). The membrane filtration tanks were also modelled by reactors with phase separation and biological activity (Figure 2.14).

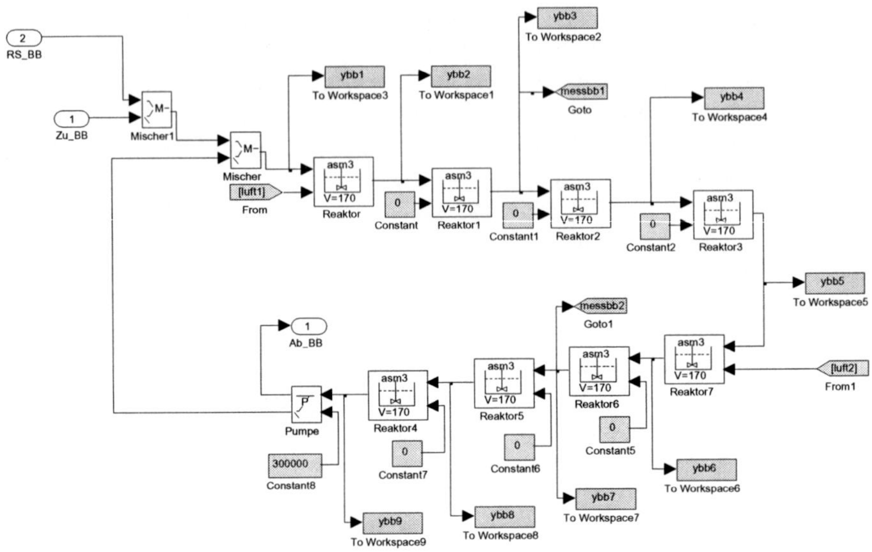

Figure 2.13 Model of the bioreactor of the Glessen MBR

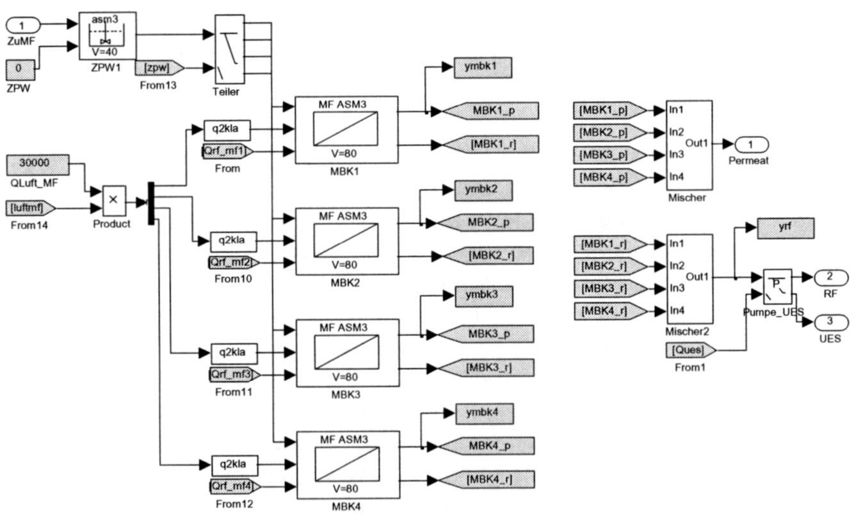

Figure 2.14 Model of the membrane filtration units of the Glessen MBR

Inflow Characterisation and Model Calibration

At Rödingen several measurement campaigns were carried out during the years 2000 to 2003 to identify the wastewater composition and the inflow variations. To assess the overall performance mass balances of influent, effluent and sludge extraction were carried out. The process dynamics were modelled based on two-hour composite samples of the relevant parameters.

Initially the Ss and Si fraction of the influent COD fractionation were identified using the methods of Ekama (1986) and Henze (1992). These values showed a wide variation and did not lead to accurate modelling results. Finally the plant was modelled and calibrated by means of a trial-and-error procedure.

The dynamic behaviour was also modelled based on 2-hour composite samples in the influent and effluent. It could be seen that the influent wastewater composition at the plant was highly dynamic. Time series of several days could not be calibrated by using a single parameter set. Therefore a dynamic influent fractionation was used distinguishing between dry and rain weather conditions. This observation can be explained by the nature of the catchment area. The catchment area is small but widely branched and the sewer has only a mild slope so that the effects of run-off, the flushing out of deposits and the dilution by the rain water are very pronounced and produce a visible effect on the plant that, compared to similar sized CAS, is aggravated by the short HRT and reduced buffering volume of the MBR. Another example of the evaluation of the COD fractions for dry and rain weather is shown in Table 2.10. An increase of both non-biodegradable components takes place during rain-events while the amount of readily biodegradeable COD is reduced.

Table 2.10 Results of the COD fractionation in the influent of Rödingen MBR of the calibration period 18.09.2000–15.10.2000 (Erftverband, 2001)

% of COD	Dry weather	Rain weather
X_I	20	40
S_I	8	15
Ss	10	1

The influent characterisation of the Rödingen model was also based on a trial and error method and calibrated on excess sludge removal over two different periods (57 and 28 days respectively) (Figure 2.15), the MLSS in the bioreactor and effluent variations of the actual plant.

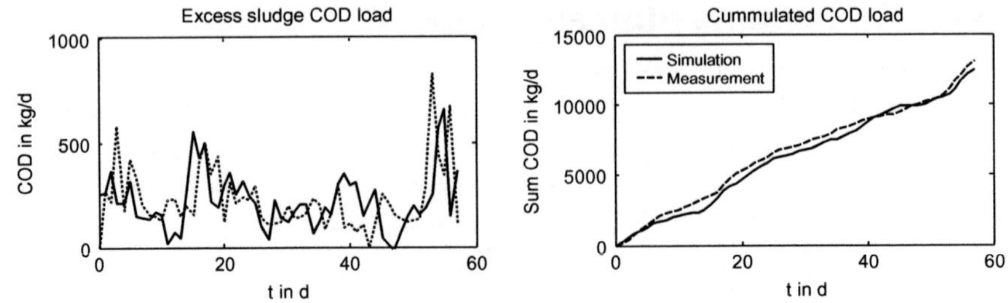

Figure 2.15 Excess sludge removal at Rödingen MBR, COD load

At the Glessen WWTP a model of the existing CAS was set-up and calibrated first using ASM2d. These data were then used to also to model the later MBR. As the SRT of the CAS and the designed MBR were similar, it was assumed that there would be no significant variation in the kinetic parameter set (DHV, 2006). Table 2.11 shows the parameters that were modified compared to the original values of a later version of the model using ASM3.

Table 2.11 Modified ASM3 model parameters of the Glessen MBR

Parameter	Description	Value
Y_{H_aer}	Yield coefficient for heterotrophs in aerobic growth [g COD/g COD]	0.84
f_{XI}	Fraction of inert COD generated in biomass lysis [g COD/g COD]	0.26
b_H	Rate constant for lysis and decay [1/d]	0.1
μ_{AUT}	Maximum growth rate of XAUT [1/d]	1.0
KN_{NH4}	Saturation coefficient for ammonium (substrate), autotrophic growth [g N/m^3]	0.1

Results

The modelling results generally show a good coincidence between real and simulated ammonia effluent peaks. As an example the ammonia and nitrate effluent concentration of Nordkanal MBR can be seen in Figure 2.16. The deviations in absolute concentration values for single events are within the accuracy of the model assumptions. Also the position of nitrate effluent peaks can be predicted while the model underestimates the denitrification capacity of the real installation. A possible explanation lies in the fact that during low inflow single filter lines are shut down to adjust the filtration capacity. Measurements have shown that during these times oxygen concentrations in the tanks become unevenly distributed. Simultaneous denitrification may thus take place in the nitrification tank.

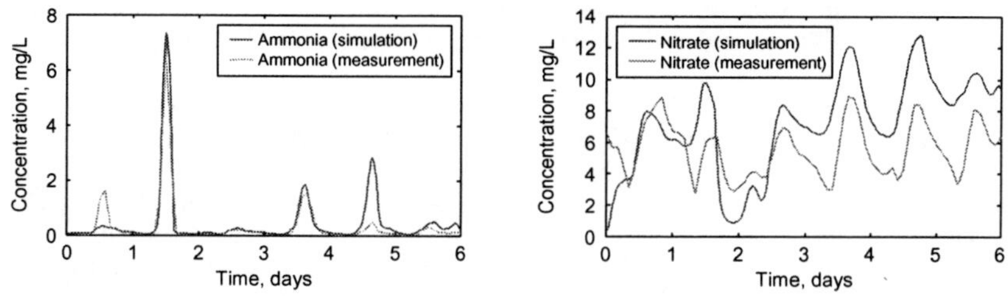

Figure 2.16 Example of simulation results of ammonia (left) and nitrate (right) concentrations, Nordkanal MBR

It can be concluded that the ASM-based models can be used to model MBRs at least as long as the SRT of the MBRs does not significantly exceed values that are also common at CAS.

REFERENCES

Brannock M. W. D., De Wever H., Wang Y. and Leslie G. (2007a). Evaluation of membrane bioreactor performance via computational fluid dynamics modelling: effect of membrane configuration and mixing. *International Conference on Membranes for Water and Wastewater Treatment*, 15–17 May, Harrogate, UK.

Brannock M. W. D., De Wever H., Wang Y. and Leslie G. (2007b). Computational fluid dynamics simulations of MBRs: inside submerged versus outside submerged membranes. *6th International Membrane Science and Technology Conference IMSTEC07*, 5–9 November 2007, Sydney, Australia.

Bornemann K., Freund M., Londong J., Nowak O., Otterpohl R. and Rolfs Th. (1998). Hinweise zur dynamischen Simulation von Belebungsanlagen mit dem Belebtschlammmodell Nr. 1 der IWAQ (Considerations on the dynamic simulation of activated sludge plants using the ASM1). Korrespondenz Abwasser, **45**(3), 445–468.

Brepols C., Schäfer, H. and Engelhardt N. (2005). Hinweise zur verfahrenstechnischen Integration getauchter Membranfilter in kommunalen Membranbelebungsanlagen. *KA – Abwasser, Abfall*, **52**(1), 45–50.

De Wever H., Brepols C. and Lesjean B. (2009). Decision tree for full-scale submerged MBR configurations. *Final MBR-Network Workshop on Salient Outcomes of the European Projects on MBR Technology*, 31 March–1 April 2009, Berlin, Germany.

DHV Water BV (2006). *Dynamic Simulation and Optimisation, Glessen WWTP.* Internal report to Erftverband, DHV Water BV, Amersfoort.

Ekama G. A. and Marais G. R. (1986). Procedures for determining influent COD fractions and the maximum specific growth rate of heterotrophs in activated sludge systems. *Water Sci. Technol.*, **18**(6), 63–89.

Erftverband (2001). *Weitergehende Optimierung einer Belebungsanlage mit Membranfiltration – Zwischenbericht (Advanced Optimisation of an Activated Sludge Plant with Membrane Filtration – Mid-Term-Report).* Report to the Ministry of Environment in North Rhine – Westpahlia, Germany, Erftverband, Bergheim, pp. 73–93.

Erftverband (2004). *Optimierung einer Belebungsanlage mit Membranfiltration* (*Optimisation of an Activated Sludge Plant with Membrane Filtration*) Report to MUNLV NW, Erftverband, Bergheim.

Fenu A., Guglielmi G., Jimenez J., Spèrandio M., Saroj D., Lesjean B., Brepols C., Thoeye C. and Nopens I. (2010). Activated sludge model (ASM) based modelling of membrane bioreactor (MBR) processes: a critical review with special regard to MBR specifitites. *Water Res*, June. (accepted for publication), *Water Res.*, **44**(15), 4272–4294.

Gujer W., Henze M., Mino T. and van Loosdrecht M. (1998). *The Activated Sludge Model No.3, IAWQ Task Group on mathematical Modelling for Design and Operation of Biological Wastewater Treatment*, IAWQ, London.

Henze M. (1992). Characterization of wastewater for modelling of activated sludge systems. *Water Sci. and Technol.*, **18**(6), 91–114.

Holthausen E. (2003). *Strömungssimulation Belebungsbecken* (Hydraulic Simulation of the Bioreactor). Internal Report to the Erftverband.

Hunze M. and Schumacher S. (2005). *Die Membranbelebung der Kläranlage Nordkanal – CFD-Modellierung – Ein Tool zur Analyse der Systemverhältnisse* (Nordkanal MBR – CFD-modelling – a tool to analyse the system conditions), 6. Aachener Tagung Siedlungswasserwirtschaft und Verfahrenstechnik, Aachen, 25–26 October 2005.

IFAK. http://simba.ifak.eu/simba/, 'SIMBA – Simulation of wastewater systems' (accessed 14 June 2010).

Pinnekamp, J. and Friedrich H. (ed) (2006). *Municipal Water and Waste Management: Membrane Technology for Waste Water Treatment.* FIW Verlag, Aachen, Germany.

Chapter 3

Performance of the wastewater treatment process

Christoph Brepols, Andreas Janot, Kinga Drensla and Torsten Engels

PRETREATMENT

Mechanical pretreatment of municipal wastewater is considered a well established technology. Screens, sieves and grit chambers thus seldom receive much attention in CAS. In an MBR, wastewater pretreatment is an intrinsic process to maintain a sustainable and safe operation of the membrane filtration units. The requirements on efficiency and operational safety of the mechanical pretreatment in an MBR differ widely from CAS applications.

In CAS plants usually the screening can be relatively coarse. In the case of the Erftverband's CAS the grid width is regularly larger than 6mm. After passing this initial influent screen and following the sand and grit chamber the wastewater still contains high amounts of substances that are of no or little influence in a CAS process, but are unwanted in an MBR: hairs and fibrous materials form braids on the membrane fibres on other parts of the modules or membrane filtration cassettes. Plastic particles, twigs or plant seeds can impair the membrane surface or block the aerator holes of the air scouring devices. Fatty or floating substances may accumulate on the filters header.

With CAS a bypass to the screening often exists to allow plant operation even during breakdowns of the screening equipment. This is a strategy that has to be considered highly critical with MBRs.

The general phenomenon of braid formation is not a matter of the membrane type used although it might be conceded that one membrane module type may be less sensitive than another. As membranes form a physical barrier for suspended solids, including hairs and fibres, it is obvious that the concentration of hairs and fibres will rise within the system if the removal rate via excess sludge is lower than the load rate, and in this case it is just a question of time before braid formation will start. Thus, the removal of hairs and fibres by screening and sieving prior to the biology is an essential task (Frechen *et al.*, 2008). After the start-up of the Rödingen MBR, the filtration modules encountered massive braid formation within a five months period of operation (Figure 3.1).

Figure 3.1 Braid formation at the Rödingen MBR, a new ZW 500 A cassette (left) and the cassette after five months of operation (right) (1999)

Once the accumulation of materials in the membrane filtration modules starts, it can turn into a self accelerating process: The module hydraulics are hampered, stagnant zones enlarge and more materials are deposited. Finally whole areas of the membrane surface are blocked with sludge and coarse materials. As a consequence, the flux in other parts of the filters increases involuntarily.

From the operation of the early Rödingen MBR, the Erftverband has bitterly learned, that an insufficient pretreatment may lead to a major breakdown of the whole filtration process within only a few months. The importance of the pretreatment was underestimated when starting up the plant in spring 1999.

The installation of two 3mm step screen devices in parallel (to avoid an unintentional bypass) seemed sufficient, as this was much greater than with the previous CAS. Anecdotal evidence from MBR applications abroad also pointed in the same direction. Although a steady decline in permeability was recorded, the resulting braid formation and membrane clogging went unnoticed until a sharp drop in wastewater temperature, the result of a sudden snow fall and snow melt in late autumn 1999, which brought the hydraulic membrane performance to its knees.

It was possible to relieve the situation through membrane cleaning, adjustments to the module hydraulic, increased air flow and sludge recycling and a reduced MLSS concentration. Later, the mechanical wastewater treatment was retrofitted by adding a rotary cutter to the inflow and a side-stream biomass screening with a 0.7mm drum screen to the process. Finally, the reconstruction of a major sewer in the catchment area reduced the amount of infiltration water and thus allowed for a decreased average flux at dry weather and longer relaxation periods of the membrane filters.

At the later MBRs at Nordkanal and Glessen, additional fine sieves were added to the mechanical pretreatment downstream of the aerated sand and grit chamber. But even the employment of 0.5 mm bar screening did not remove enough hair and fibrous materials to control the braiding and clogging of the modules entirely. The sieves at the Nordkanal MBR thus underwent several retrofits before finally a 1.5mm wire mesh sieve was installed in 2010. Figure 3.2 shows different types of sieve geometry that were employed at the Nordkanal MBR. The photos make the bearing structure of the sieves visible. The free surface of the geometries is different. For improved cleaning an additional high pressure nozzle was installed with the mesh sieve, as can be seen in the righthand picture.

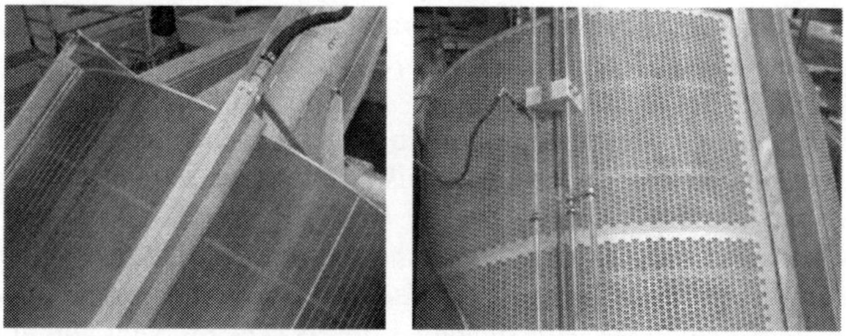

Figure 3.2 Nordkanal MBR, drum sieve with 0.5mm bar/slit geometry (left) and 1.0mm mesh geometry (right)

The experience of the Erftverband has shown that sieve hydraulics, sieve cleaning and removal of materials require intensive optimisation to reduce the number of operational breakdowns as well as operational cost. As the sieve cleaning is mostly performed by using permeate it may eventually consume a significant share of the produced permeate and thus become a relevant cost factor. The non optimised sieve cleaning that resulted in very short cleaning intervals at the Nordkanal MBR could eventually require up to 10% of the total amount of permeate produced with dry-weather inflow.

Schier *et al.* (2009a) also emphasised the influence of the screen and sieve geometries on the removal efficiency and the hydraulic capacity of this treatment step. The same authors also emphasise that not only the sieve but also the combination of treatment steps has to be scrutinised holistically while the employment of an additional sieve does not necessarily relieve the operational disruptions inn an MBR plant but partly shifts their focus from the membrane filtration units to the pretreatment (Schier *et al.*, 2009b).

AERATION AND OXYGEN SUPPLY

Aeration is an important parameter in the design and operation of an MBR. It is required for the biological treatment, the biomass agitation and membrane scouring. While different systems are employed in the biological compartment, the air scouring of the membrane filters is achieved with the built-in aerator tubes. Although the main purpose of the air scouring is membrane cleaning, it may also contribute significantly to the oxygen supply of the biomass and hence influence the operation of the biological compartment of the MBR.

Process Aeration

Table 3.1 gives an overview of the different aerator types used for the biological process aeration. The specific oxygen transfer rate (SOTR) was measured on site using the method according to DWA M209.

In 2002/2003 the membrane diffuser at Rödingen had to be changed from EPDM to silicone membranes due to biologically induced material deterioration. At that time, similar observations with this diffuser type were also made at CAS plants, so this cannot be considered to be MBR specific. The new silicone diffusers provided only about 95% of the oxygen transfer that was observed with the EPDM diffusers (Erftverband, 2004). However, it was decided also to equip the Nordkanal MBR with silicone diffusers. The problem with the EPDM diffusers was later sorted out by the supplier, such that today EPDM is used again for the replacement of old or defective membrane diffusers.

Table 3.1 Process aeration and specific oxygen transfer rate (SOTR), biological compartment

Plant	Rödingen	Glessen	Nordkanal
Aeration	Fine bubble aeration	Surface aeration	Fine bubble aeration
Aerator type	Membrane plate diffusers	Brush, 800mm diameter	Membrane plate diffusers
Material	EPDM	–	Silicone
Aerator depth	4.0m	–	4.0m
Aerator length	–	2 × 4.0m	–
SOTR (water)	$18.7gO_2m^{-3}m^{-1}$	–	$15.6gO_2m^{-3}m^{-1}$
αSOTR (biomass)	$8.2gO_2m^{-3}m^{-1}$	$4.2–5.2kgO_2h^{-1}m^{-1}$	$8.3gO_2m^{-3}m^{-1}$
$MLSS_{bio}$	14g/L	8g/L	12g/L

Air Scouring

The hollow fibre membrane filters at the three MBRs require intermittent air scouring with coarse bubbles for surface cleaning. The aeration is an integrated part of the membrane module design. Originally the Rödingen MBR relied on permanent air scouring. Todays intermittent mode of operation was introduced in 2001 as a means of energy reduction. The aeration rate is set according to the membrane suppliers recommendations for a sustainable operation of the membrane filters. The specific aeration rate is approximately $0.4m^3m^{-2}h^{-1}$ regardless of the actual permeate production. At all three plants one blower supplies one filtration line. While the membrane filtration line is in operation the respective blower operates constantly. The intermittent aeration is achieved by channelling the air to different parts of a filter line. The reference value of the aeration is 10s on/10s off. Lag time of the shifting valves and the control equipment may actually lead to up to 50% longer aeration cycles.

At Rödingen, the specific oxygen transfer rate SOTR for the air scouring in the filtration compartment was $10gO_2m^{-3}m^{-1}$ in plain water, while α SOTR was between $3–5gO_2m^{-3}m^{-1}$ in biomass. At Nordkanal SOTR of the air scouring was $6.15gO_2m^{-3}m^{-1}$ in plain water and $3.2gO_2m^{-3}m^{-1}$ under process conditions. This might suggest that the difference between inside configuration as at Nordkanal or in separate filtration compartments as at Rödingen could be responsible for the observed difference.

Aeration Efficiency and MLSS

The aerator efficiency can be correlated with the MLSS concentration in the bioreactor although the MLSS concentration is not by itself the limiting factor but influences the rheology of the liquid. The gas transfer is, among many other influencing factors, also influenced by the viscosity.

At Rödingen systematic test trials with different MLSS concentrations have been carried out. The alpha factor as the coefficient of aerator efficiency with plain water and under process conditions is displayed in Figure 3.3. It could be seen also that the alpha-factor is not a constant that is solely correlated to MLSS concentration or sludge viscosity. In practice, the alpha factor is subject to daily variations as are the properties of the activated sludge. The alpha factor decreases with increasing MLSS concentration. It was found that the alpha factor in the filtration compartment tended to be lower at comparable MLSS concentrations (Erftverband, 2004).

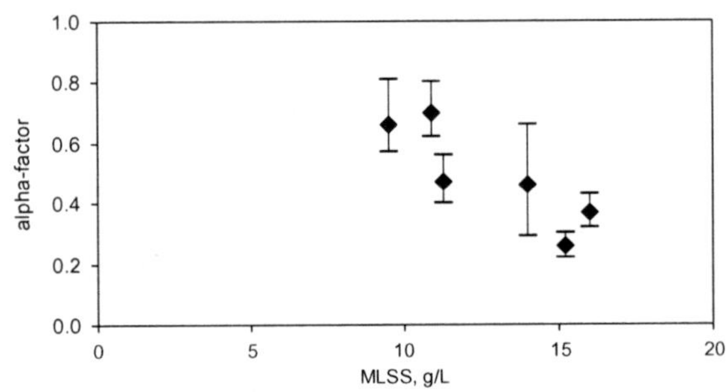

Figure 3.3 Alpha-factor and MLSS concentration in the bioreactor at the Rödingen MBR (Erftverband, 2004)

Despite the possible bandwidth of values that can be found, Krause (2005) published an approximate function of the α-factor versus the MLSS concentration of $\alpha = e^{-0.056MLSS}$ for the design of municipal MBRs. In this function α equals 0.5 for an MLSS concentration of 12g/L.

NUTRIENT ELIMINATION

Actual Efficiency of the Biological Treatment

With respect to their SRT and F:M ratio the three MBRs are operated under comparable conditions as conventional contact stabilisation plants. However, the HRT is significantly shorter because of the reduced bioreactor volume and the lack of a secondary clarifier. Although the short HRT leads to a more pronounced dynamic of the biological treatment process MBRs are found to be capable of maintaining even strict effluent regulations under peak conditions.

The process configuration of the three MBRs is different as is the on-site equipment for monitoring the influent and effluent quality. At Rödingen only the effluent quality of ammonia, nitrate and phosphate is monitored via online sensors. The Nordkanal MBR employs online influent and effluent quality measurements for the values of ammonia, nitrate, phosphorous and SAC-UV. The Glessen plant uses influent measurements of ammonia and phosphate and effluent measurements of ammonia, nitrate and phosphate. At all three plants, additional values are analysed on a regular basis from mixed and grab samples.

Rödingen MBR. The Rödingen MBR is designed for full nitrification at a minimum wastewater temperature of 5°C. There are no effluent regulations for nitrate. In effect the denitrification is at least partly omitted at low wastewater temperatures in order to increase the nitrification capacity.

Since 2001 the plant has been operated with intermittent denitrification. Because of strong seasonal variations, the operation in pre-denitrification mode with a fixed aerobic to anoxic volume ratio of 0.5 in the bioreactor was not flexible enough to guarantee the required ammonia effluent concentrations throughout the year. Figure 3.4 shows the effluent concentrations of ammonia and nitrate that were measured from 2 hour composite samples and grab samples of the permeate between 2001 and 2007. Ammonia effluent concentrations of more than 0.5mg/L are only encountered occasionally. Nevertheless, for 60% of the time, a degree of denitrification can be achieved, which would be sufficient even under more stringent regulations.

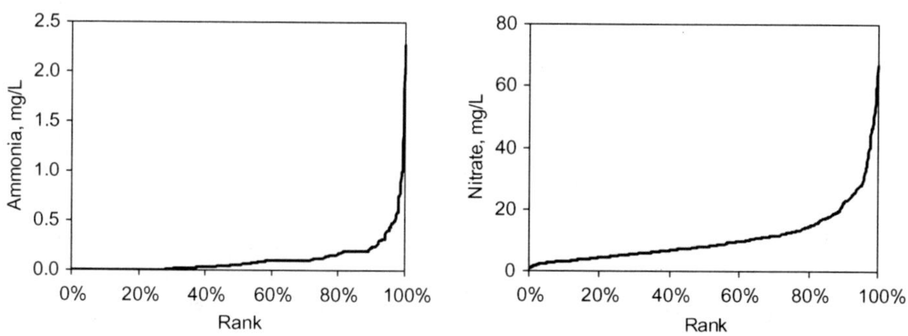

Figure 3.4 The relative frequency of occurrence of the effluent concentrations of ammonia (left) and nitrate (right) at the Rödingen MBR 2001 to 2007

Glessen MBR. The different treatment efficiencies of a CAS without tertiary treatment and an MBR can be illustrated by the example of the start-up of the retrofitted MBR at Glessen. Figure 3.5 shows the effluent concentrations of COD, ammonia and phosphate before and after commissioning the membrane filtration units. The operation was different only in that the secondary clarifier was replaced by the separate membrane filtration. Other operational settings were left unchanged at first. The COD and phosphate values immediately decreased due to the complete retention of any suspended particles from the effluent. Nitrification also takes place in the filtration compartments, where oxygen is abundant while permeate is produced. As a result, the number and size of ammonia effluent peaks decreased although the nitrification/denitrification did not improve thoroughly until the new process control was fully operative and optimized. On-line measurements of ammonia in the bioreactor and in the permeate show how the ammonia concentration is reduced in the filtration tanks. The variations in Figure 3.6 show the two-hour average values during a week with dry weather conditions, the influent concentration of total nitrogen averages 47mg/L. The effluent concentration from the bioreactor is further reduced to 0.1mg/L in the filtration compartments. The low effluent concentrations had implications for the process control of the nitrification/denitrification as is explained below.

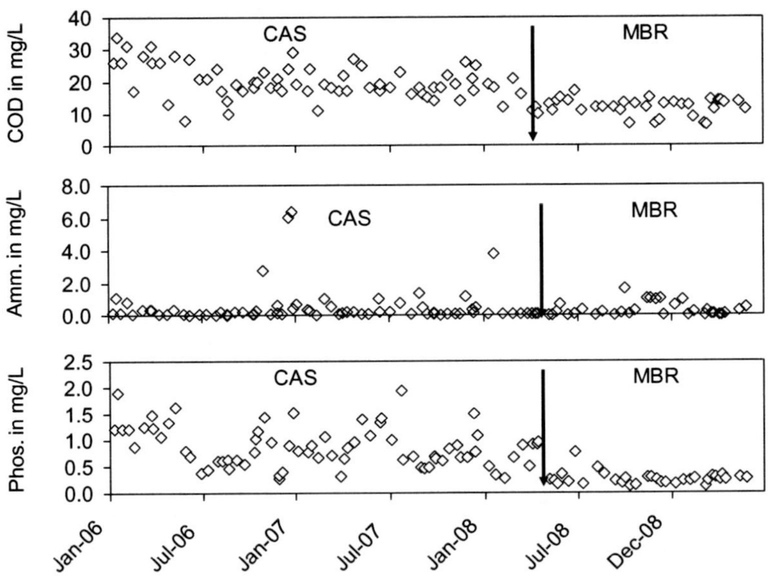

Figure 3.5 Effluent concentrations of COD (above), ammonia (middle) and phosphorous (below) at the Glessen WWTP before and after the commissioning the MBR

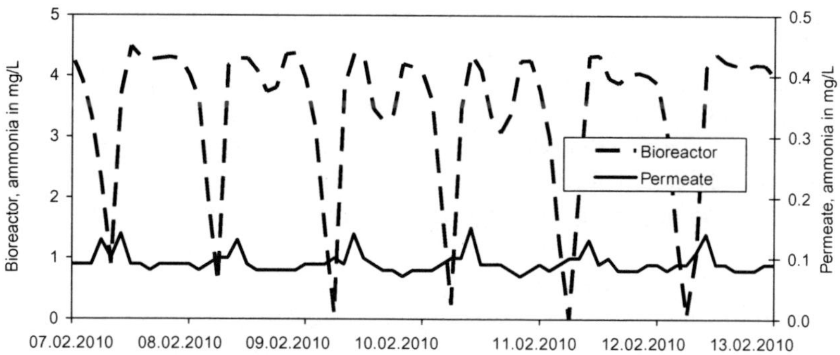

Figure 3.6 Daily variations of the ammonia concentration in the bioreactor and the permeate at the Glessen MBR, two hour average values from 07.02.2010–13.02.2010

Nordkanal MBR. The different data sources that are available at the Nordkanal plant have been used to assess the performance of the biological treatment processes of the MBR. The average concentrations and loads are used to calculate the actual denitrification rate of the plant. The daily loading of chemical oxygen demand COD, total nitrogen TN and phosphorous P_{tot} for the respective period can be seen from Table 3.2. The quantile $Q_{.85}$ of the influent loading accounts for more than 90% of the plants design capacity. The biomass content is an average MLSS of 12g/L while volatile suspended solids (VSS) are at 7g/L. The F:M ratio averages 0.05gCOD/gMLSS. An average denitrification rate of 1gN $gVSS^{-1}$ h^{-1} and a denitrification efficiency of approximately 85% can be calculated.

Table 3.2 Daily loading (quantile $Q_{.85}$ and average values of sample data) Nordkanal WWTP

Daily loading	COD	TN	Ammonia	Nitrate	P_{tot}
Influent ($Q_{.85}$/avg.), kg/d	9000/6000	810/570	620/450	–	150/80
Effluent ($Q_{.85}$/avg.), kg/d	400/230	160/80	4/1	130/60	6/4
Sludge production, kg/d	4500	265			145
Design influent, kg/d	9600	897	–	–	123

Figure 3.7 shows the daily maxima and average effluent concentrations of ammonia as recorded by the installed on-line sensors during the period from 01.01.2007 to 31.12.2007. The Average ammonia concentrations are 40mg/L in the influent and 0.12mg/L in the effluent. Despite these low effluent values single peaks of up to 9mg/L occur during the winter season at low temperatures

when nitrification capacity is at its minimum. The gradient of the curve suggests that denitrification performance is relatively stable throughout the year. Nitrate effluent concentrations are at times lower than originally expected. At times of low inflow the nitrification compartment is only partially aerated. Oxygen concentrations that were then measured in different parts of the reactor suggest that anoxic conditions may exist, so that denitrification may take place temporarily also in parts of the original nitrification basins.

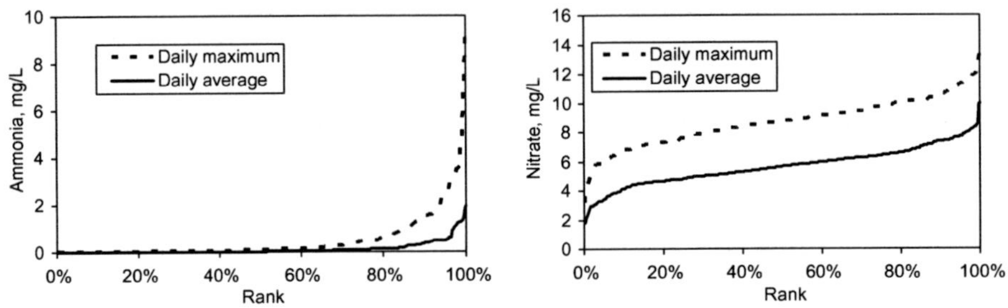

Figure 3.7 The relative frequency of occurrence of the daily average and maximum effluent concentrations of ammonia (left) and nitrate (right) at the Nordkanal MBR, 01.01.2007 to 31.12.2007

The influent and the effluent concentrations show considerable variations especially under rain weather flow. Figure 3.8 gives an impression of the variability of the influent load of ammonia and the respective effluent concentrations.

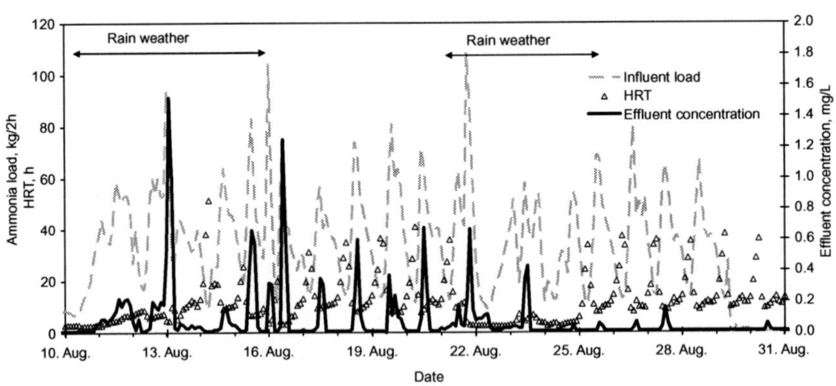

Figure 3.8 Example of influent and effluent variations of ammonia and HRT at Nordkanal MBR

The time variation curves of the on-line measurements in the influent and effluent reveal, that the effluent ammonia peaks occur not later than four hours after the influent peak is measured. The exact time lag between these incidents depends of the inflow and the respective HRT. A more detailed investigation of selected dry weather and rain weather events shows that the time lag is significantly shorter than the average HRT as can be seen from Table 3.3.

Table 3.3 Time lag between influent and effluent ammonia peaks

Inflow	HRT, h	Time lag T, h	Relative time lag, T/HRT
Dry weather	>9.0	2.0–4.0	<0.30
Rain weather	4.7–6.0	0.5–1.5	0.10–0.25

A comparison of the observed relative time lag under maximum rain weather conditions to the theoretical values of the residence time under peak flow (also Chapter 2) validates the characteristics of a completely mixed reactor.

The practical experience at Nordkanal shows that the required effluent quality can be maintained under all circumstances, although the actual hydraulic residence time is considerably low.

Process Control Strategies

Oxygen supply is controlled automatically at all three MBRs. The usual setting of 1 to $2mgO_2/L$ is increased when the ammonia effluent concentration reaches a defined threshold. With pre-denitrification, which today is only in use in the Nordkanal MBR, the nitrate recycle can either be controlled according to the inflow, the nitrate concentration in the effluent or both. The intermittent denitrification is time-controlled by the duration of the anoxic and aerobic phases. As in the Glessen MBR the duration can be offset by defined ammonia and nitrate set points. Figure 3.9 gives an example of a possible control loop for a pre-denitrification plant. A cascade controller is used to set the oxygen concentration in the bioreactor.

Efficiency of the Denitrification Process

One of the main impediments for denitrification in MBRs is the oxygen–carryover from the filtration compartment. The air scouring in the membrane filtration compartments leads to a high oxygen concentration. This oxygen load reaches the bioreactor with the return sludge. When the return sludge is fed directly to the denitrification tank the oxygen content may disturb the anoxic ambient that is required for denitrification. On the other hand, a reduction of the return sludge rate reduces the possible MLSS concentration in the bioreactor (also Chapter 2). As Figure 3.10 shows, the oxygen carryover increases proportionally to the return sludge rate.

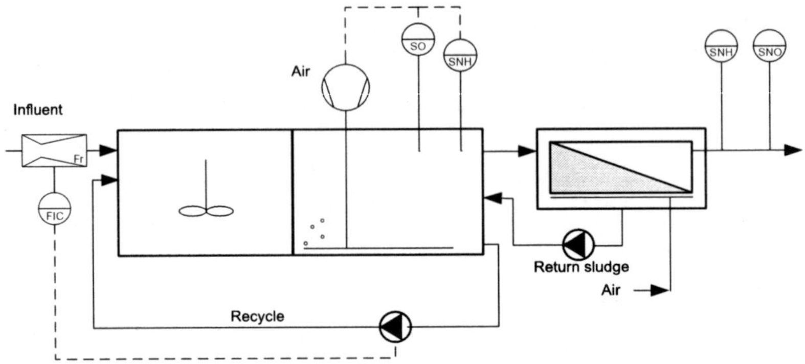

Figure 3.9 Example of a process control scheme for an MBR with pre-denitrification

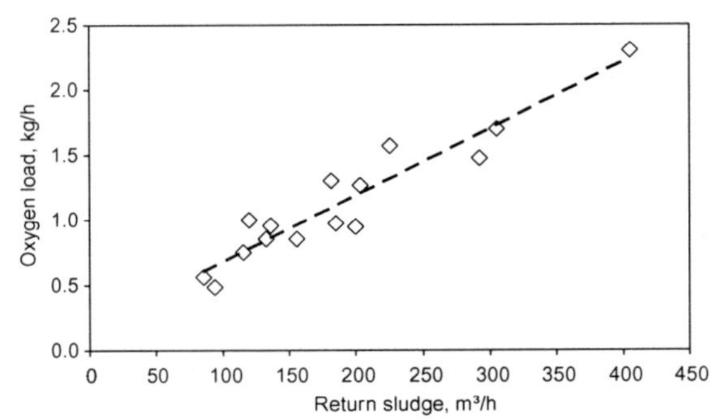

Figure 3.10 Recycled oxygen load at the Rödingen MBR (Erftverband, 2004)

As a consequence, the trade-off between the achievable MLSS concentration and the oxygen carryover is responsible for a reduced denitrification capacity. The benefit of an increased nitrogen recycling and a higher biomass concentration in the denitrification compartment of the MBR is partially offset by the increasing oxygen carryover. As a consequence, the operational range to achieve a sufficient denitrification is smaller when compared to CAS with pre-denitrification.

Small WWTPs like the Rödingen and Glessen MBR usually suffer from more pronounced variations of their influent load than large installations. A flexible operational regime is thus required when low effluent concentrations are attained.

Intermittent aeration is flexible in achieving nitrification/denitrification by adjusting the aeration cycles according to operating conditions like effluent concentrations or temperature. However, it always requires separate filtration tanks as the filtration cannot be operated under anoxic conditions due to the obligatory air scouring. The control loop must facilitate a quick response to changing boundary conditions.

In the Glessen MBR the duration of the respective nitrification and denitrification steps are controlled according to the effluent concentrations of ammonia and nitrate. Originally, an automatic process photometer for the determination of ammonium in the permeate was in use to retrieve a feed-back signal for the control loop that then starts aeration or changes the oxygen set point in the bioreactor. At normal dry weather conditions the average ammonia concentration in the permeate is very low and variations are only benign (Figure 3.6). Sample preparation times of the device and the hydraulic retention time in the filtration tanks were responsible for a delay of more than 30 to 60 minutes before a critical increase of the effluent concentration could be detected. In combination with the slow increase of the oxygen supply achieved by the surface aerators, ammonia effluent peaks could pass-by the bioreactor before arriving at a sufficient oxygen concentration. This was due to the control loop being too slow. As a consequence, aerobic conditions had to be sustained over long periods at the cost of a low denitrification rate. In June 2009, the control loop was optimized by using an ion-selective probe to determine the ammonia concentration directly inside the bioreactor. This reduced the time-delay significantly. The set-points of the control loop could thus be adjusted in favour of the denitrification process while ammonia effluent peaks were diminished successfully. The values in Figure 3.11 show that the total nitrogen effluent could be reduced from approximately 7 to 3mg/L without a significant increase in the ammonia average.

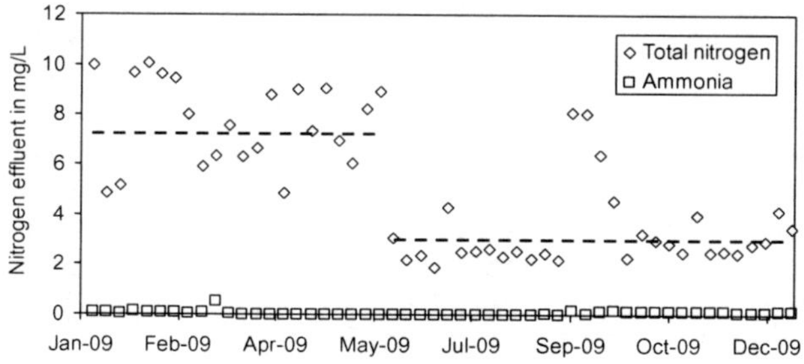

Figure 3.11 Nitrogen effluent concentrations at Glessen MBR before and after optimizing the process control loop

PHOSPHOROUS REMOVAL

All three MBRs are operated with simultaneous phosphorous precipitation. The precipitate is either dosed to the return sludge where there are separate filtration tanks (Rödingen and Glessen) or into the bioreactor inflow where there is a combined bioreactor/filtration tank (Nordkanal).

Ferric chloride or ferric chloride and ferrous sulphate are the common precipitants. At CAS aluminates plus polymer are sometimes used during the cold season in order to prevent bulking sludge. The use of aluminates as precipitant was also tested at Nordkanal MBR. The use of this precipitant led to more pronounced variations of the effluent concentration and difficulties in maintaining the required effluent concentration. Therefore the test was abandoned after a few weeks.

The long-term phosphorous removal efficiency of the Rödingen and Nordkanal MBR has been compared with six CAS. Four CAS are equipped with separate anaerobic reactors for enhanced biological phosphorous removal. The results in Table 3.4 are average values of the influent and effluent load and the removal efficiency. All plants show relatively high removal rates and average effluent concentrations lower that 1.0mg/L. The ratio of precipitant used versus the phosphorous load in the influent can be seen as an indicator of the dominating removal mechanism. High values indicate the chemical phosphorous precipitation with salts of iron or aluminium is dominant while lower values hint at an increasing dominance of the biological phosphorous removal. The molar Me/P-ratio shows a bandwidth between 0.22 and 1.30. The values of the MBRs are more in the range of the CAS with an anaerobic reactor.

Table 3.4 Phosphorous removal efficiency of various municipal WWTPs

Plant type	Anaerobic tank	P influent load, kg/d	P effluent, mg/L	Removal rate	Me/P-ratio
MBR	No	5.9	0.23	98%	0.53
CAS	No	9.2	0.93	84%	0.99
CAS	Yes	24.0	0.23	94%	0.90
CAS	Yes	26.0	0.46	93%	0.54
CAS	No	49.7	0.43	95%	1.30
MBR	No	145.5	0.29	97%	0.58
CAS	Yes	145.7	0.46	97%	0.87
CAS	Yes	268.1	0.31	93%	0.22

Silva *et al.* (2009) found relatively high numbers of phosphorous accumulating bacteria (PAO) in several full scale MBRs (including Nordkanal

MBR). A fluorescence in situ hybridisation (FISH) technique was used to identify PAO and glycogen accumulating organisms (GAO) in the activated sludge. The levels were similar for MBR plants with and without a dedicated anaerobic compartment. It was assumed, that anaerobic micro-niches inside the biomass flocs could be responsible for these observations as the flocs show a more dense and compact structure than flocs from normal CAS.

It has to be acknowledged, that trimming down the phosphorous effluent concentrations by massive overdosing of precipitants may increase inorganic fouling on the membrane surface and ferrous residues inside the fibres and permeate pipes may interfere with the chemical membrane cleaning.

DISINFECTION

At the Glessen MBR the microbial quality of the effluent was tested before and after the start-up of the membrane filtration. The results in Figure 3.12 show the immediate effect of the membrane filtration as a physical barrier.

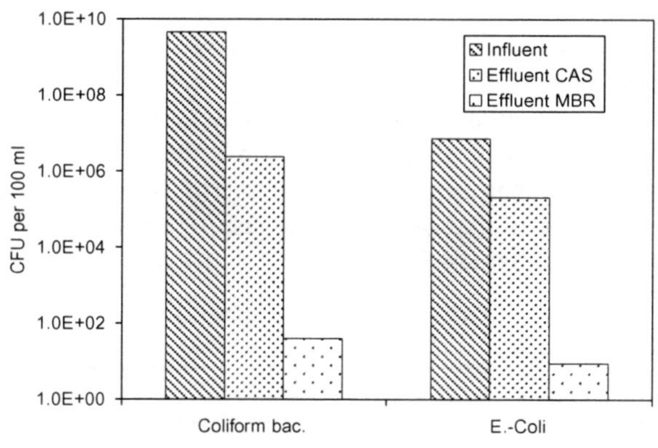

Figure 3.12 Logarithmic values of influent and effluent concentrations of coliform bacteria and escherichia coli (E-coli) at the Glessen WWTP before an after retrofitting the plant as an MBR

The number of the microorganisms in the effluent of the Nordkanal MBR was sampled during the years 2006 to 2008. Grab samples were taken from the mixed effluent of all 8 filtration lines, results are displayed in Table 3.5. The sampling point is an open channel, thus the influence of external microbial pollution cannot be fully excluded. Generally a good disinfection quality is achieved. A clear trend towards a degradation of the disinfection properties with time could to date not be observed.

Table 3.5 Results of the determining colony counts, Nordkanal MBR

Sampling date	CFU at 20°C ml^{-1}	CFU at 36°C ml^{-1}	E.coli 100 ml^{-1}	Colifo-rmous 100 ml^{-1}	Salmo-nella 100 ml^{-1}	Entero-coccus 100 ml^{-1}
31.01.2006	258	7	0	2	0	0
11.04.2006	46	2	2	4	0	0
26.07.2006	82	94	0	1	0	0
24.04.2007	57	8	1	15	0	0
04.07.2007	118	174	2	10	0	0
29.04.2008	8	7	0	1	0	1
04.08.2008	2	60	0	4	0	0
29.10.2008	33	50	0	17	0	0

SLUDGE AND WASTE PRODUCTION

Screening Waste

The amount of debris that is processed at the screens increases disproportionally with the decrease of the screen bar width. After commissioning the Nordkanal MBR in 2004, the cumulative amount of debris which is produced at an upstream pumping station in the sewer (20mm), the 6mm influent step screen and the 0.5 mm drum bar screen downstream of the aerated sand and grid chamber. The amount of debris increases disproportionately to the decrease of the bar-screen width (Figure 3.13).

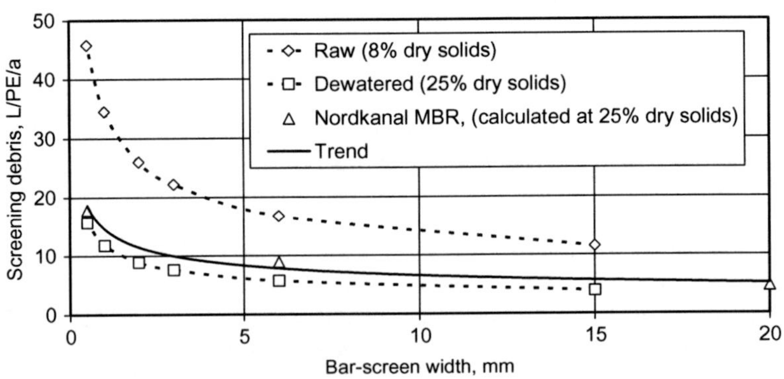

Figure 3.13 Amount of screening debris at Nordkanal MBR, compared to values of Seyfried *et al.* (1985)

After changing the sieve geometries from a bar/slit geometry to a wire mesh, the quantity of debris increased dramatically. At times the amount tripled, leading to a shortage in storage capacity for the dewatered material and to regular overflows of the storage container. The situation could only be relieved with the processing of this screening waste together with the biological surplus sludge. The provisional equipment, which can be seen in Figure 3.14, was later replaced by a new automatically controlled installation.

Figure 3.14 Nordkanal MBR, container overflow (left) and provisional equipment for feeding the diluted screening debris to the sludge treatment (right)

Biological Surplus Sludge Production and Dewatering

The operational values that were used to calculate the specific sludge production are displayed in Table 3.6. The overall average sludge production at Rödingen MBR is 0.48kgMLSS/kgCOD. The average SRT is, at 38 days, significantly higher than the intended design value. At Glessen the average sludge production is 0.61kgMLSS/kgCOD at an SRT of 33 days. Sludge dewatering at both plants is off-site.

At the Nordkanal MBR the overall average sludge production is 0.55kgMLSS/kgCOD. The average SRT is 25 days. Sludge dewatering is on-site using a centrifuge.

Figure 3.15 shows the biological surplus sludge production versus the annual WWTP inflow. The sludge production at the Nordkanal MBR apparently exceeds the sludge production of similar sized CAS plants. When extrapolating the trend-line for WWTPs without primary sedimentation it can be found that the Nordkanal data are close to that range. However, the Nordkanal MBR is the only plant with a capacity of more than 2,000,000m^3 per year that does not employ primary settling. The organic loading in the bioreactor influent is thus approximately 30–40% higher than that of the comparable CAS. Sludge production from primary settling is not considered in the diagrams.

Table 3.6 Operational values, sludge production

Average values	Unit	Rödingen	Glessen	Nordkanal
Daily inflow	m^3/d	540	2100	15000
Total COD in	mg/L	680	380	580
Total Nitrogen in	mg/L	72	47	52
Ammonia in	mg/L	50	30	40
BOD/COD ratio		0.7	0.5	0.4
COD out	mg/L	18.9	12.0	18.4
Ammonia out	mg/L	0.2	0.3	0.4
TN out	mg/L	13	8	5
Temperature	°C	15	14	15
$MLSS_{bio}$	mg/L	11.5	8.0	11.9
Sludge production	kg/d	120	490	4400

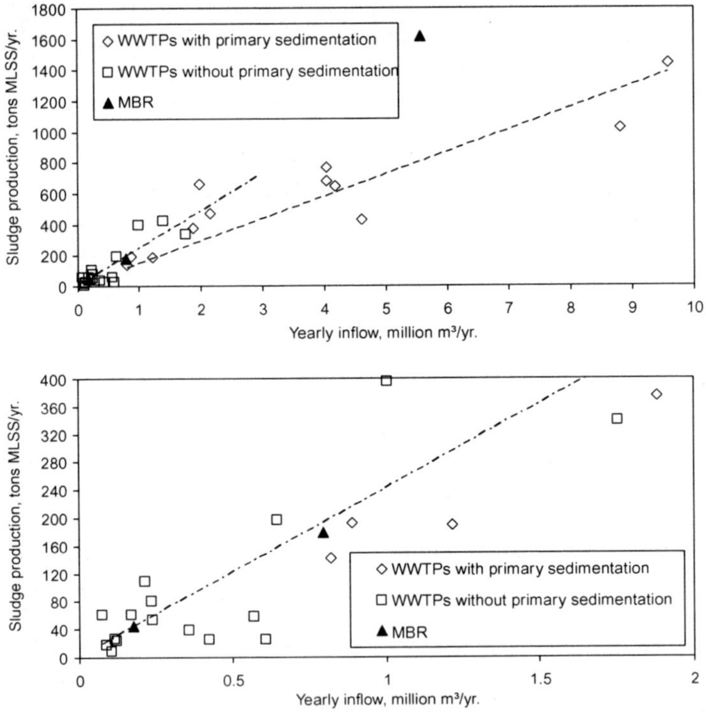

Figure 3.15 Biological surplus sludge production at WWTPs in the Erft region with and without primary sedimentation, year 2009

The dewatering results of 17 WWTPs have been compared on the basis of a long-term series of grab samples from the dewatered sludge that where taken at least twice a year and were analyzed at the Erftverbands central laboratories. Sludge treatment at the Erftverband is semi-centralized. Sludge from smaller installations is stored on site and periodically transported to larger plants, where it is further treated or dewatered. The sludge is eventually mixed with sludge from other installations and dewatered using centrifuges, chamber filter presses or belt presses. Among the plants with on-site dewatering 4 plants, including the Nordkanal MBR, employ simultaneous aerobic sludge stabilization in the bioreactor. All other 13 plants use anaerobic digesters for sludge treatment. Due to the treatment concept the dewatered sludge on these sites may sometimes be a mixture of sludge from anaerobic digesters and contact stabilization plants. Table 3.7 shows the process configurations of the WWTPs.

Table 3.7 Process configuration of the sludge treatment of the WWTPs

Number	Primary settling	Sludge digester	Dewatering unit	Stationary
1	no	no	chamber filter press	no
2	no	no	chamber filter press	no
3	no	no	chamber filter press	no
4	no	no	centrifuge	yes
5	yes	yes	centrifuge	yes
6	yes	yes	centrifuge	yes
7	yes	yes	belt press	yes
8	yes	yes	centrifuge	yes
9	no	yes	centrifuge	yes
10	yes	yes	centrifuge	yes
11	yes	yes	chamber filter press	yes
12	yes	yes	centrifuge	yes
13	yes	yes	centrifuge	yes
14	yes	yes	belt press	yes
15	yes	yes	centrifuge	yes
16	yes	yes	centrifuge	yes
17	yes	yes	centrifuge	yes

Figure 3.16 shows the range of the dry solids content of the dewatered sludge. As expected, the sludge from anaerobic digesters in general reaches a higher dry solids content than that from contact stabilization plants. The dewatering results of the Nordkanal MBR sludge are similar to other aerobically stabilized sludge.

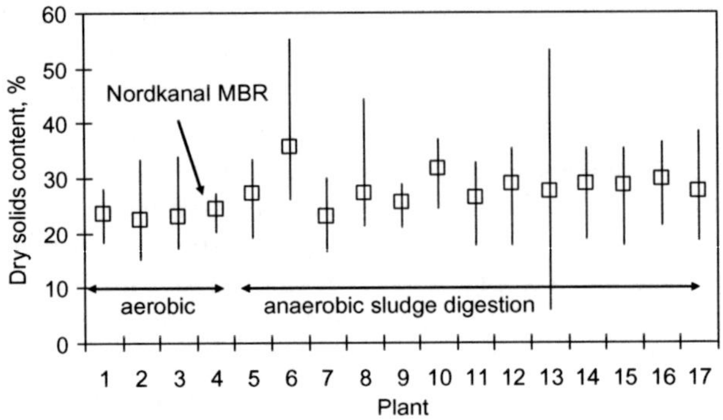

Figure 3.16 Average, minimum and maximum values of the dry solids content of the dewatered sludge at the Erftverband's WWTPs with simultaneous aerobic sludge stabilisation and separate anaerobic digesters employing on-site sludge dewatering

Combined Dewatering of Surplus Sludge and Fine Screen Debris

The sludge dewatering results at Nordkanal were improved by remixing the debris from the influent fine screening to the surplus sludge. In test trials, the maximum MLSS concentration in the dewatered sludge could be augmented from 25% up to 30–35%. These values are not obtained in normal operation, nevertheless the overall amount of dewatered sludge could be decreased from an average of 14.2m^3/d in 2005 and 2006 to a present average value of 13.5m^3/d.

REFERENCES

DWA M209. (2007). Merkblatt 209, Messung der Sauerstoffzufuhr von Belüftungseinrichtungen in Belebungsanlagen in Reinwasser und in belebtem Schlamm (Measurement of the Oxygen Transfer in Activated Sludge Aeration Tanks with Clean Water and in Mixed Liquor). Deutsche Vereinigung für Wasserwirtschaft, Abwasser und Abfall e. V., Hennef, Germany.

Erftverband (2004). *Optimierung einer Belebungsanlage mit Membranfiltration*. Report to MUNLV NW, Erftverband, Bergheim, Germany.

Frechen F. B., Schier W. and Linden C. (2008). Pre-treatment of municipal MBR applications. *Desalination*, **231**(1–3), 108–114.

Krause S. (2005). Untersuchungen zum Energiebedarf von Membranbelebungsanlagen (Research into the energy consumption of membrane bioreactor plants). Schriftenreihe WAR 166, Darmstadt, Techn. Universität, Diss., 2005.

Schier W., Frechen, F.-B. and Fischer St. (2009a). Efficiency of mechanical pretreatment on European MBR plants. *Desalination*, **236**(1–3), 85–93.

Schier W., Frechen F.-B., Drensla K., Janot A., Engelhardt N., Exler H. and Ohme M. (2009b). Ganzheitlicher Blick auf die Abwasservorbehandlung (MAV) auf kommunalen Membranbelebungsanlagen (Holistic view at mechanical pretreatment on municipal MBR plants). Begleitbuch zur 8. Aachener Tagung Wasser und Membranen, 27–28. Oktober 2009.

Seyfried S. and Lohse B. (1985). Vergleich der Reinigungsleistungen von Rechen, Sieben und Siebrechen sowie deren Einfluß auf die weiteren Reinigungsstufen (Comparison of the removal efficiency of screens, sieves and rake sieves and their influence on the subsequent treatment steps), Schriftenreihe Veröffentlichungen des Institutes für Siedlungswasserwirtschaft und Abfalltechnik der Leibniz-Universität Hannover; 58, Hannover, 1985.

Silva A. F., Carvalho G., Lousada Ferreira M., Nieuwenhuijzen van A., Guglielmi G., Crespo J. G., Reis M. A. M. and Crespo M. T. B. (2009). Microbial population structure of pilot and full scale membrane bioreactors. *Proceedings of Final MBR Network Workshop*, 31 March–1 April, Berlin, Germany.

Chapter 4

Effects on the water quality in the receiving waters

Heinrich Schäfer

INTRODUCTION

Due to the anthropological influences such as coal mining, agriculture, fisheries, use of water power and flood protection, many bodies of water have been seriously affected in the past decades in comparison with their natural status. The water quality is also heavily impaired through residential drainage. The targets defined in the 1990s in Europe and achieved in many cases for the improvement of the water quality, have been updated in 2000 with the EU Water Framework Directive. According to this, the surface waters are, within a defined time frame, to be transformed into the so-called good condition. So far as this is not to be achieved with a body of water as the result of strong change, the so-called good ecological potential is to be realised. In the same way the good condition is also demanded for bodies of groundwater.

IMMISSION-ORIENTED OBJECTIVES OF THE TREATMENT OF WASTEWATER

The advantages of membrane technology, in addition to the small use of area and the high adaptability, are in the excellent treatment performance. The very

 Operating Large Scale Membrane Bioreactors for Municipal Wastewater Treatment. By Christoph Brepols. ISBN: 9781843393054. Published by IWA Publishing, London, UK.

extensive carbon and nutrient removal achievable using conventional wastewater engineering is improved even further using membrane technology. The effluent is quasi free of solid matter. The obvious difference, however, lies in the very extensive disinfection of the wastewater. On the other hand, the elimination of pharmaceutical products, trace elements or pesticides in MBR plants continues to be a topic of the discussion.

The requirement on the treatment performance of a wastewater treatment plant is dependent on the following immission-oriented objectives and constraints:

(1) Water pollution control, i.e. objective of water quality, influenced inter alia by:
   a. hydrological situation, i.e. relative proportions of body of water and discharge
   b. preloading of the body of water, inter alia through other discharges
   c. structure of the body of water
   d. repopulation potential of the body of water

(2) Groundwater (resources) protection:
   a. separation between body of water aquifer
   b. soil conditions
   c. filtration effects of the soil

(3) Other usage-oriented objectives:
   a. use of bank filtrate
   b. use of the body of water for bathing
   c. drawing of water for irrigation
   d. direct use of treated wastewater for irrigation

In addition to the immission-oriented objectives the type of drinking water processing is of significance with a use of raw water.

Within the scope of the monitoring for the implementation of the Water Resources Directive a multitude of assessment parameters (macrozoobenthos, structure of the body of water, saprobes, fish fauna, macrophytes, phytobenthos, diatoms, phytoplankton, chemical parameters and priority substances) have been developed and applied for the first time. Analysis data over many years are comprehensively available with the Erftverband, however, only for saprobes for the assessment of the water quality. With the saprobes system the relationships between organism communities in the body of flowing water and their pollution are drawn on. The saprobic index (Zelinka, Marvan, 1961) is then the defining factor of the resulting representation, which takes place in quality classes between I (no loading – good quality) to IV (excessive loading) and their intermediate steps.

EFFECTS OF THE RÖDINGEN MBR ON THE FINKELBACH

The Rödingen wastewater treatment plant drains into the Finkelbach, a tributary of the Erft. The Finkelbach runs in a west–east direction and is considered to be part of the Börden low land system of bodies of water with rurally characterised environment made up of arable land or grassland. Due to the location in the direct catchment area of the brown coal open cast mining a groundwater level remote from the surface is produced. It is fed solely by a precipitation-conditioned surface runoff and discharges from the residential drainage system. As a result of the lack of native groundwater inflows the Finkelbach is an ephemerous body of water, i.e. it dry becomes temporarily where it is not continuously fed with water from a wastewater treatment plant (Figure 4.1).

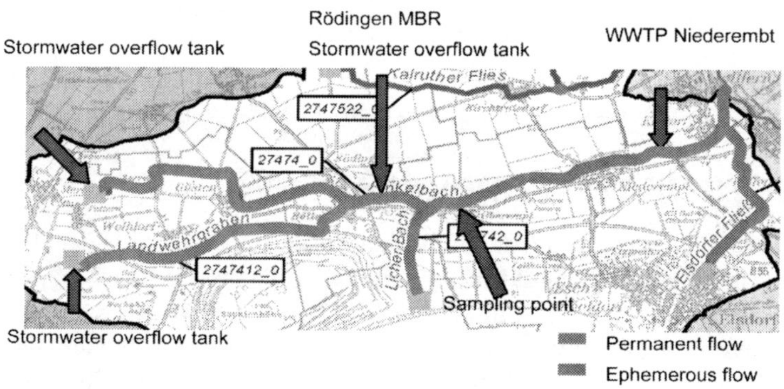

Figure 4.1 Catchment area of the Finkelbach stream

Objective of the Rödingen WWTP is Water Quality Class II (or II–III). Due to the quantitative dominance of the wastewater treatment plant discharge this demands the most extensive removal of nitrogen and phosphorus in the WWTP. The Finkelbach is not a body of bathing water, a contact with the aquifer quasi does not exist so that disinfection is also not required.

As expected, the water quality has, continuously improved since the commissioning of the Rödingen MBR, so that today the Finkelbach possesses Water Quality Class II–III (Figure 4.2).

For the necessary improvement of the water quality stormwater tanks with overflow and the retention soil filters have been established for the treatment of precipitation water. This sets in motion the general requirement of an overall consideration of the pollution sources. Also, a further improvement of the water quality can at best only be achieved through reductions of the loading from other, for example diffuse, sources.

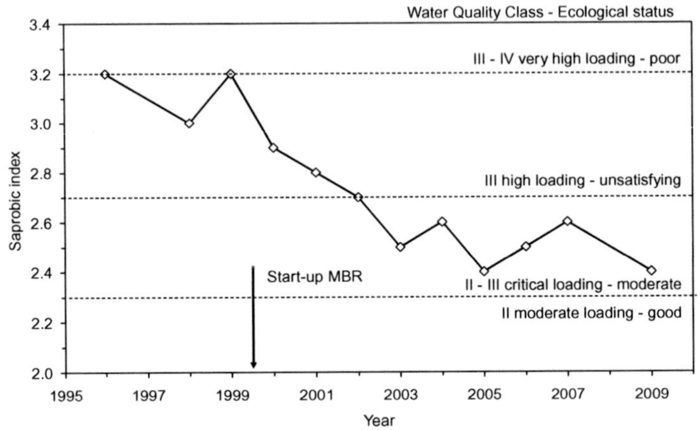

Figure 4.2 Development of water quality in the Finkelbach between 1996 and 2009

EFFECTS OF THE NORDKANAL MBR

The Nordkanal is the remnant of the unfinished Napoleonic 'canal du nord' between the rivers Rhine and Maas (Figure 4.3). Today the Nordkanal runs over only 18km between the cities of Willich and Neuss in a west–east direction and is an extremely slow flowing stream. Another stream, the Jüchener Bach, flows into the headwater of the Nordkanal. This stream obtains its small water discharge through sump water discharges and from the effluent of the Erftverbands conventional Glehn WWTP. The Nordkanal itself flows into a branch of the Erft near Neuss, which ultimately pours into the inner harbour in Neuss.

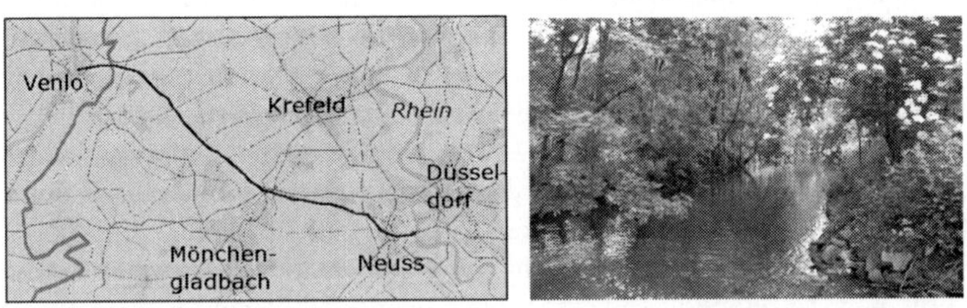

Figure 4.3 Map of the intended original course of the Nordkanal waterway (left) and the Nordkanal today between Willich and Neuss (right)

Analogous to Rödingen, the immission-oriented objective for the Nordkanal is Water Quality Class II (or II–III). For the necessary reduction of the discharged loads, the Erftverband has established the Nordkanal MBR. Since commissioning in 2004 the water quality has improved by more than one class (Figure 4.4), so that today Water Quality Class II is achieved. Analogous to the Finkelbach, a hygienic improvement of the Nordkanal is probable but, due to the lack of analysis data, not verified.

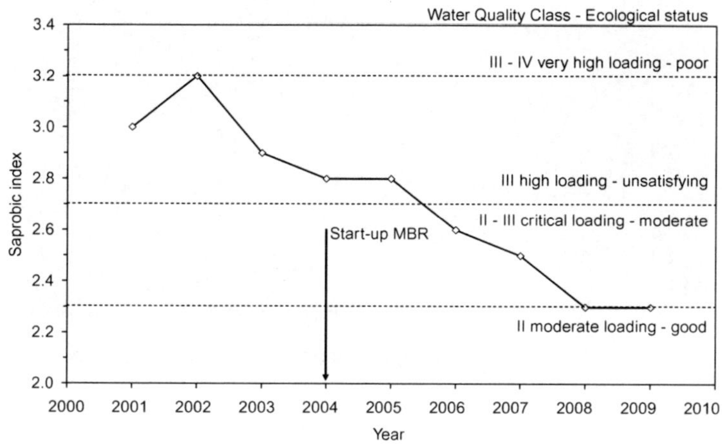

Figure 4.4 Development of water quality in the Nordkanal between 2001 and 2009

EFFECTS OF THE GLESSEN MBR

The Glessen MBR discharges into a small stream, the Glessener Bach. The water quality of the Glessener Bach was already good before the retrofitting of the Glessen WWTP. As described in Chapter 2, the main objective of the retrofitting with MBR technology was the improved groundwater protection in the catchment area of the neighbouring Weiler waterworks. The Glessener Bach percolates in a wetland contributing to the aquifer (Figure 4.5).

In the past no influence of the discharge from the Glessen WWTP on the raw water in Weiler has been determined, as pathogens are completely removed, as a result of the long flow time with the soil passage. However, based on the principle of precaution an MBR was demanded, in that already with the wastewater treatment, an extensive removal of pathogens is achieved as part of a multibarrier system. (Lindner *et al.*, 2004). The actual effects of the membrane filtration following commissioning of the MBR plant in 2008 are still not detected and are to be confirmed in later analysis programmes.

Figure 4.5 View of the wetland 'Große Laache' fed by the Glessener Bach

REFERENCES

Lindner W., Engelhardt N., Rose U. and Schäfer H. (2004). Wasserversorgung, Abwasserentsorgung, Biotop- und Ressourcenschutz im Widerspruch? Erfordernis einer ganzheitlichen Betrachtung des Feuchtgebietes Große Laache (Water supply, sanitation, habitat and resource conservation in conflict? Requirement of a holistic view on the wetland Große Laache). *GWF Wasser Abwasser*, **145**(11), 782–787.

Zelinka M. and Marvan P. (1961). Zur Präzisierung der biologischen Klassifikation der Reinheit fließender Gewässer (To the specification of the biological classification of the purity of flowing waters). *Arch. Hydrobiol.*, **57**, 389–407.

Chapter 5

Long term membrane performance

Kinga Drensla, Christoph Brepols, Michael Trimborn and Andreas Janot

INTRODUCTION

All three MBRs in the Erftverband employ GE Zenon ZeeWeed hollow fibre membranes. In 1999 the Rödingen MBR was originally equipped with ZW 500 A membrane cassettes. After severe operational problems due to insufficient wastewater pretreatment it was decided to replace these with new ZW 500 C cassettes. The actual replacement took place in April 2001 for the first and April 2003 for the second line. At that time the tendering of the Nordkanal MBR had already brought the decision, that membranes of the ZW 500 C type would also be installed at this plant. The replacement at Rödingen thus was also seen as a form of pilot using the new membrane type. At Glessen MBR, the public tendering again resulted in GE Zenon membranes. In 2008 new ZW 500 D modules were installed there. It was only recently, that one of the two filtration lines of the Rödingen MBR was retrofitted, this time with KMS Puron modules.

When starting up the first MBR in April 1999, the effort involved in cleaning the hollow fibre membranes was not yet apparent. However, a sharp decline in membrane performance during the first winter season made it obvious that cleaning and maintenance of the filters are crucial and intricate processes in MBR operation.

These initial problems with coarse fouling and clogging of the membranes were overcome by optimizing the filter tank and module configuration and upgrading the mechanical pretreatment of the sewage. The average concentration of suspended solids in the filtration tanks was decreased. Later a retrofit of the municipal sewer system reduced the infiltration of surface waters to the main sewer. As a result, the average inflow to the plant decreased reducing also the average hydraulic load of the filtration.

In the following years, the Erftverband optimized the chemical cleaning methods at the Rödingen MBR, trying various cleaning agents and procedures. These trials, mainly carried out on a large scale at the plant, were performed within the context of a research project (Erftverband, 2004). The resulting optimized methods were already adopted in the new Nordkanal MBR during planning and commissioning phases and later also adopted in the Glessen MBR.

HYDRAULIC PERFORMANCE

The hydraulic performance of a membrane filter as the amount of permeate produced over time is influenced by a variety of factors. Some of them are pre-defined by the set-up of the filter modules and the design of the MBR plant while others can be influenced by the plant operation. Unfortunately it is not always possible to quantify their impact. This is even more true for large plants like the Nordkanal MBR (Figure 5.1) where scale effects interfere with the basic process characteristics.

Figure 5.1 Permeate suction pipes and aeration pipes of a filtration line (left) and permeate pumps (right), Nordkanal MBR

The key elements of any membrane process relate to the influence of the following parameters on the overall permeate flux: (a) membrane resistance, (b) the operational driving force per unit membrane area, (c) the hydrodynamic conditions at the membrane/liquid interface, (d) the fouling and subsequent cleaning of the membrane surface (Judd, 2006).

The following list gives an overview of influencing factors that play a role in practical plant operation.

1. Wastewater properties
   - Temperature
   - Content of organic and inorganic compounds
2. Inflow characteristics
   - Inflow variations
   - Frequency and duration of peak inflow conditions
3. Performance of the pretreatment
   - Screening
   - Sand and grit chambers
   - Sieves
   - Primary clarifiers
4. Design and operation of the bioreactor
   - Biomass fouling
   - MLSS concentration
   - SRT
   - HRT
   - F/M-ratio
   - Reactor configuration
   - Biological nutrient removal
5. Membrane properties and operation
   - Module hydraulics
   - Filtration tank layout
   - Average and maximum flux
   - Trans-membrane pressure (TMP)
   - Chemical resistance
   - Backwashing regime
   - Relaxation
   - Cleaning strategy
   - Air scouring

Operational practice has to aim at achieving a sustainable operation of the hollow fibre membrane filtration over many years. Besides the boundary conditions from the plant design, such as MLSS concentrations or membrane flux, the regular monitoring of the variations in membrane performance, membrane cleaning and membrane maintenance help to ensure successful operation.

Definition of Permeability as a Key Parameter

The hydraulic capacity can be defined by the flux as the specific flow per unit membrane area, the trans-membrane pressure (TMP) as the driving force and the permeability L_p.

$$L_p = \frac{Q}{A_M \cdot \Delta p_{TM}}$$

L_p: Permeability in $L/(m^2\ h\ bar)$
Q: Flow in L/h
A_M: Membrane area in m^2
Δp_{TM}: Trans-membrane pressure (TMP) in bar

In practice the unit of permeability is often noted as LMH/bar.

Permeability is a global parameter, that summarizes different processes. The permeability is inversely related to the filtration resistance and, in a model description, can be seen as a serial connection of single resistances (Wintgens, 2003).

$$L_p = 1 / [(R_M + R_{CL} + R_F) . \eta]$$

R_M: Membrane resistance
R_{CL}: Cake layer resistance
R_F: Fouling resistance
η: Viscosity

Permeability Measurement

There is no generally accepted procedure to determine permeability in an MBR. The determination of permeability values on a full scale installation is sensitive to various factors. As a result, values can differ widely from plant to plant or operator to operator, even if the same types of membrane filters are used. In order to monitor the variations of permeability over a long period it is necessary to stay with a certain protocol that defines the boundary conditions at the specific site.

As a derived quantity the accurate determination of the permeability requires the measurement definition of three items: (1) flow, (2) the TMP that are measured on-site and (3) the membrane area, which is defined by the installed membrane filtration units in a line. With the submerged membrane filtration units used at the MBRs TMP has only very low absolute values. Practically it has to be determined as the difference between the static pressure at the pressure gauge while the filtration is in pause mode and the dynamic pressure while the filtration is running. Inaccuracies in the measurement of these two values lead to a misdiagnose of permeability.

At the Erftverband, a manual determination of permeability values of each filtration line at the MBRs is carried out weekly under controlled boundary conditions according to the following protocol:

(1) the membrane filters are backwashed,
(2) the water level in the filtration tank is set to its maximum at the beginning of the measurement,
(3) the static pressure at the pressure gauge is determined,
(4) the flow is set manually to its maximum design value,
(5) filtration in started,
(6) the dynamic pressure is read when its value becomes nearly constant.

Complementary to this manual method, permeability is determined automatically at the Nordkanal and Glessen MBR based on on-line measurements of the required values. For the automatic determination and documentation in the process control system it is important, that the underlying algorithm is case-sensitive for filtration and backwashing, as otherwise no reasonable average data are calculated.

Aside from malfunctioning pressure gauges and flow meters the following inaccuracies occurred at the Rödingen MBR (Erftverband, 2004) and can also be found on the other installations (Table 5.1).

Table 5.1 Influence of the boundary conditions on the determination of the TMP

Cause	Effect	TMP
Gas accumulation in the pipe works	Additional hydraulic loss	to low
Mistimed reading	Working pressure is not yet constant	to low
Previous high flow	TMP should be determined after backwashing or a relaxation period	to low/high
Remaining vacuum in permeate pipe	Wrong static pressure	to low
Incorrect flow adjustment	TMP and flux rise disproportionately due to dynamic hydraulic loss in the piping	to low/high
Water level in filtration tank	Influence on static pressure	to low/high

In full-scale MBR plants, often large numbers of membrane filtration modules operate in one line. When flux and TMP are measured, the result is always only an average value for the whole line. The local conditions with single modules within a line or even with single membrane fibres may differ significantly.

Monitoring the variations of membrane performance is an indicator for the long-term behaviour of the filtration. For this purpose permeability is often given at a standardised temperature (e.g. 20°C).

The Erftverband uses different compensation factors, that are approximating the slope of the dynamic viscosity of water versus temperature using the equation $L_{p,T=20°C} = L_{p,T} \cdot k^{(20-T)}$. At Nordkanal MBR for $T < 20°C$, $k = 1.033$ and for $T > 20°C$, $k = 1.025$. At Rödingen and Glessen MBR a slightly different value is used with $k = 1.022$ (Figure 5.2). For reasons of coherence with older data these different values are still in use. Practically this is of little significance only.

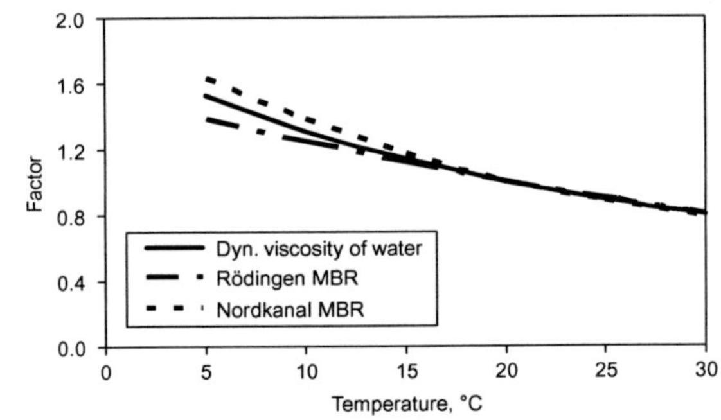

Figure 5.2 Temperature compensation of permeability at 20°C and normalised dynamic viscosity of water

Figure 5.3 gives the weekly average values of the permeability of all 8 filtration lines at the Nordkanal MBR. After the plant start−up permeability declined relatively quickly because of the non optimized cleaning protocol. It was only after the chemical cleaning strategy was revised in 2005 that permeability recuperated. It can also be seen that there are differences in permeability between the lines although the overall operating conditions are similar.

Figure 5.4 shows the actual values of the permeability that were determined on one filtration line compared with the calculated temperature compensation of the average permeability of 210LMH/bar at 20°C. It can be seen that the projected decline of the permeability values is slightly overestimated when using only the temperature compensation for the dynamic viscosity. Although other operational parameters such as SVI, MLSS concentration, COD removal rate or denitrification rate are also subject to seasonal variations, the main influence was probably membrane cleaning. The in-situ maintenance cleaning is performed at shorter intervals with cold temperatures and recovery cleaning was performed before the winter season.

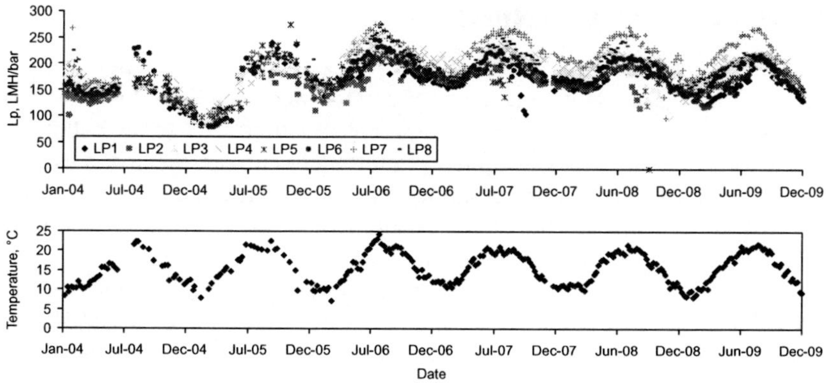

Figure 5.3 Weekly average permeability and temperature values at Nordkanal MBR 2004–2009

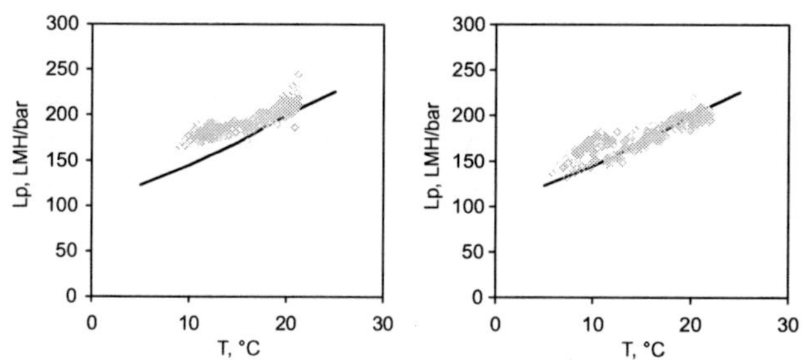

Figure 5.4 Permeability versus wastewater temperature, actual values of filtration line 3, Nordkanal MBR (dots) and temperature compensation (line) for the years 2007 (left) and 2009 (right)

MEMBRANE CLEANING AND MAINTENANCE

Four cleaning methods are employed to maintain the hydraulic performance of the submerged hollow fibre membrane filters:

Mechanical cleaning of the membranes by

(1) permanent or intermittent air scouring, and
(2) frequent, intermittent backwashing and/ or relaxation,

Chemical cleaning is performed by either

(3) backwashing with cleaning agents, as a maintenance cleaning in place (CIP) or
(4) cleaning out of place (COP), that is performed as an occasional intensive recovery cleaning.

MECHANICAL CLEANING

Air scouring is an essential part of all submerged membrane filtration units used in municipal wastewater treatment. Aerators are installed underneath the membrane filters as a part of the module construction. Coarse air bubbles rise along the membrane surface and induce a turbulent current. Shear forces reduce the cake layer on the membrane surface. In hollow fibre modules, the induced fibre movement increases the shear. At the Erftverband MBRs the air flow is adjusted according to the membrane supplier's recommendations. The specific membrane aeration demand per unit membrane area is in the range of $0.40 Nm^3\,h^{-1}\,m^{-2}$. Air scouring is the main energy consumer of the MBRs.

Reinforced hollow fibre membranes and also some flat sheet modules can be backwashed with permeate by reversing the permeate flow. The filtration process thus becomes discontinuous. (For example, the filtration lines are backwashed every 500 to 800 seconds for 30 to 50 seconds). Alternatively or additionally relaxation can be performed by allowing the filtration units to discontinue operation regularly after a filtration cycle or at times of low inflow.

The backwashing and relaxation mode influence the yield of the filtration process. Figure 5.5 shows the actual flux and TMP variations of a filtration backwash cycle at Nordkanal MBR. The suction pressure TMP during filtration is displayed as negative value. The ramp of the permeate pump has a significant influence on the actual yield when compared to theoretical considerations that assume the flux variations follow a rectangular curve.

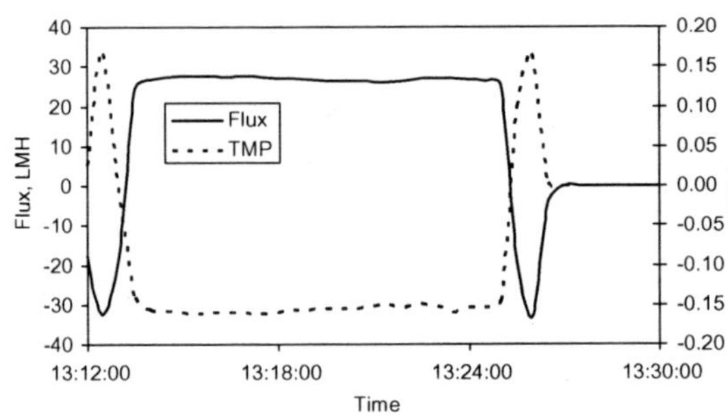

Figure 5.5 Flux and TMP variations during a filtration/backwash cycle at Nordkanal MBR

CHEMICAL CLEANING METHODS

Applying an appropriate chemical cleaning protocol is one main strategy of fouling control and amelioration at full-scale operating MBRs (Judd, 2006).

For economic reasons it is desirable to operate the same membranes over five, ten or more years. The cleaning strategy should thus help to maintain membrane performance over a long period. Besides aiming at high cleaning efficiency, a superior strategy should therefore be developed with a view towards fulfilling the following objectives:

- minimizing emission of unwanted substances in the effluent,
- being compatible with the whole treatment process,
- reducing the risk of damaging the membranes,
- minimizing plant downtime during cleaning,
- lowering labour input,
- reducing chemical cost,
- avoiding any hazards to plant operators.

To achieve these objectives, the cleaning strategy has to be composed of several individual methods. The right cleaning agents, sequences of cleaning steps and frequencies of applying each method need to be determined. Membrane and module properties, the hydraulic load, wastewater and biomass properties, wastewater pretreatment, bioreactor design and mode of operation, water temperatures and effluent regulations have also to be taken into account.

Generally the conditions at municipal MBRs change with time. The equipment ages, residues on the membrane filters or in the permeate system build up and the wastewater composition may also change. As a consequence, the cleaning protocols and strategies need to be adapted repeatedly. In view of this fact, the cleaning equipment of an MBR should ideally be designed to accommodate a wide variety of cleaning methods. Frequent CIPs are performed either in biomass or in air in an emptied filtration tank. The occasional intensive recovery cleaning can be performed externally as COP.

Table 5.2 shows typical operational values for the different methods at the Erftverband MBRs. Duration, concentrations and pH values have been optimized to the above objectives. The cleaning steps, using the different cleaning agents, are performed one after another. It is important to purge the equipment with plain permeate to wash out any residues from the cleaning before changing the cleaning agent. The pH values of the solutions are adjusted by adding low concentrations of caustic soda or HCl. The actual amount depends on the alkalinity of the permeate.

Effective de-aeration of the filters and permeate pipes was found crucial prior to the chemical cleaning (Melin *et al.*, 2007). Otherwise remaining air bubbles will hamper the distribution of cleaning agents among the modules and the filters will remain partially fouled.

The actual backwashing during the CIP takes place at intervals. A short backwash cycle is followed by a longer pause where the cleaning agents soak the membrane fibres. For example, the cleaning at Nordkanal MBR is performed using 7 cycles each of 60 seconds backwashing and 300 seconds soaking with NaOCl and NaOH, followed by a 1000 seconds backwash with plain permeate to wash out any residues of cleaning agents from the piping and the filters and then followed by 7 cycles of backwashing and soaking with citric acid an HCl.

Table 5.2 Comparison of different chemical cleaning protocols at the Erftverband MBRs

Parameter	Unit	CIP in air	CIP in biomass	COP
Alkaline cleaning NaOCl + NaOH	pH	9–10.5	9–10.5	10.5
NaOCl dosage	ppm	150–500	150–200	1,000
Acidic cleaning citric acid + HCl	pH	2	2	2
Citric acid dosage	ppm	2,000	2,000	2,000
Temperature	°C	ambient	ambient	25–35
Duration	hrs	6	2	8–24
Permeability recovery	%	0–10	0–10	50–100
Frequency	1/year	23–52	52	0.5–2
Downtime for filter line	hrs	6	2	> 24

Before a COP the membrane filters cassettes are removed from the filtration tank and are then sprayed with water to rinse off adherent biomass before inserting the membrane filters into the COP tank. The cleaning then can be performed automatically. The duration of the cleaning steps is set to optimise the workflow. Cleaning takes place within a 24 hour shift, such that the cleaned filtration cassettes in the COP tank can be replaced by a set of new cassettes to be cleaned on the following day.

Cleaning Agents

Different mechanisms of membrane blocking (Figure 5.6) require different chemical cleaning agents. In general a whole variety of cleaning agents can be used to remove or reduce membrane fouling and scaling. Table 5.3 gives an overview of the different categories of cleaning chemicals.

The effect of the acidic cleaners lies mainly in their capability to form highly soluble salts with the ions of calcium (Ca^{++}), magnesium (Mg^{++}) and iron (Fe^{+++}). The inorganic acids HCl and HNO_3 are capable of dissolving carbonates and hydroxides but not calcium sulphate ($CaSO_4$). Citric acid, because of its chelating potential, is capable of dissolving calcium compounds even in the presence of sulphate.

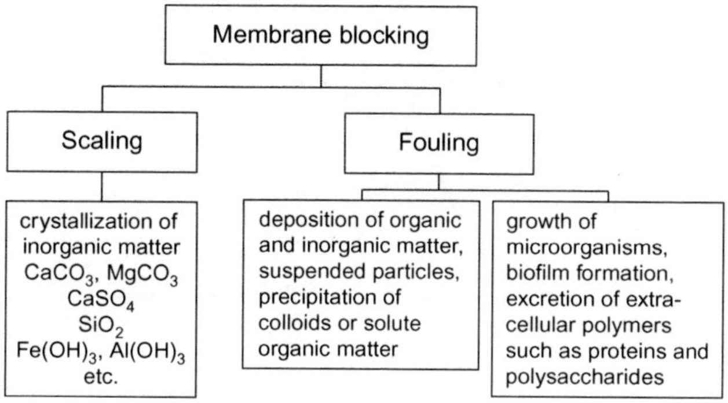

Figure 5.6 Mechanisms of membrane blocking

Table 5.3 Overview of different commercially available cleaning agents and their cleaning effects

Category	Chemical	Main effect
Acidic	Citric acid, HCl, HNO_3	Solubilisation, chelation
Alkaline	NaOH	Hydrolysis, solubilisation
Oxidative	NaOCl, H_2O_2, peroxyacetic acid	Oxidation, disinfection
Detergents	Surfactants, chelating agents, enzymes	Emulsification, dispersion, chelation, surface conditioning

Hypochlorite is used to oxidise organic compounds. Besides the oxidative effect, its application leads to chlorination of organic matter and in consequence also to the formation of unwanted by-products such as halogenated organic compounds (AOX) or chlorine gas. To control this effect it is usually used in combination with caustic soda. Because of the dissociation equilibrium of the hypochloric acid (Figure 5.7) there is only a relatively small window for proper application: High pH levels may be prohibitive as they are harmful to the membrane material itself, as stated in the suppliers manual, while at lower pH values chlorination becomes the dominating effect and thus AOX formation increases.

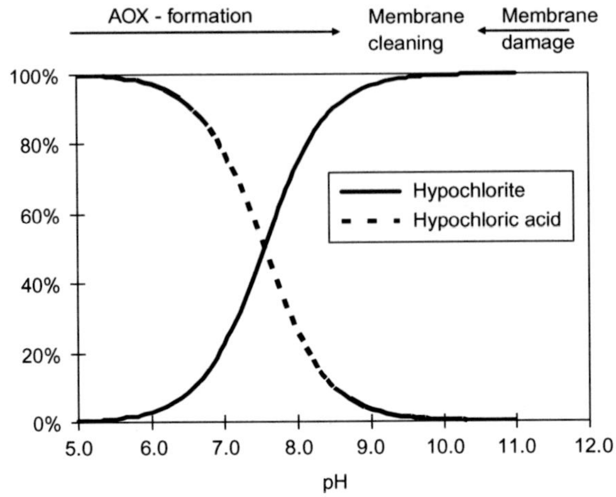

Figure 5.7 Dissociation equilibrium of hypochloric acid

Cleaning Equipment–CIP

The MBRs employ different types of equipment for the CIP as can be seen in Figures 5.8 and 5.9.

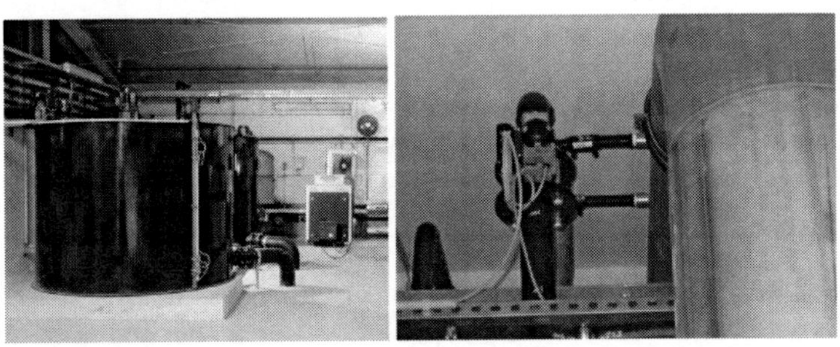

Figure 5.8 Separate CIP tanks for alkaline and acidic cleaning at Glessen MBR (left), in line dosing of cleaning agents into the permeate pipe at Nordkanal MBR (right)

The raw solution of the cleaning agents is stored in small containers with a size between 50 and 1000L depending on the size of the plant and chemical consumption. Before use, the cleaning agents have to be diluted with permeate to the required concentration. This can be done either by a step-wise dilution in the piping system or by mixing them with permeate stored in separate CIP tanks. At Rödingen MBR a single CIP tank is used for each of the two filtration lines that is also the permeate reservoir for the regular permeate backwash. At Glessen two separate CIP tanks for acidic and alkaline cleaning are used serving all four filtration lines. Additional equipment is required to lower the water level in the filtration tanks before the CIP in air can be performed.

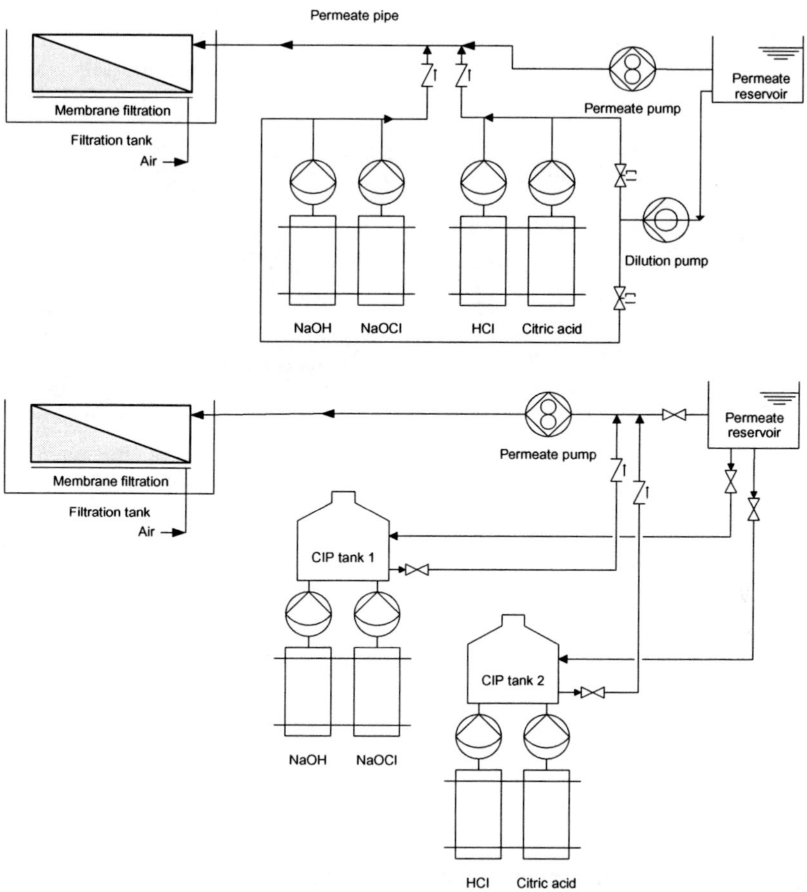

Figure 5.9 Simplified schematic of the cleaning equipment for in-situ cleaning. In-line dosing of the cleaning agents (Nordkanal MBR, above) and separate CIP tanks (Glessen MBR, below)

Cleaning Equipment–COP

COP is performed using an external device, which can be installed either in a mobile container as in Rödingen or adjacent to the membrane filtration tanks as in Nordkanal and Glessen. It consists of a small membrane filtration tank that can accommodate one or several membrane cassettes at a time. The walling of the concrete tanks is coated to achieve higher chemical resistance. A separate blower for air-scouring and a permeate/backwash pump are installed to perform the cleaning. Additional dosing facilities, storage tanks and heating devices are used to process and heat the cleaning solution. The membrane cassettes have to be inserted manually, using a crane (Figure 5.10).

Figure 5.10 Inserting a ZW 500 D cassette into the COP tank at Glessen MBR

A key feature of the COP equipment at Nordkanal MBR are the blue twin storage tanks for acidic and alkaline cleaning agent solution (Figure 5.11) with a volume of $60m^3$ each. Heat exchangers allow pre-heating of the cleaning solution to up to 35°C. The used cleaning solution is fed back to the storage tanks. Solids and debris are allowed to sediment at the bottom and are then withdrawn and processed separately. A large part of the cleaning solution and its thermal energy can thus be recovered. The amount of cleaning waste and the required heating energy are reduced. Figure 5.12 shows a simplified schematic diagram of a COP device. At Nordkanal, 10 ZW 500 C cassettes can be cleaned at a time in two COP tanks.

PRACTICAL OPTIMISATION OF THE CHEMICAL CLEANING METHODS

At the Rödingen MBR initially sodium hypochlorite and caustic soda were used extensively for backwashing and cleaning the membrane filters. This led

Figure 5.11 Storage tanks for the ready-to-use cleaning solution, Nordkanal MBR

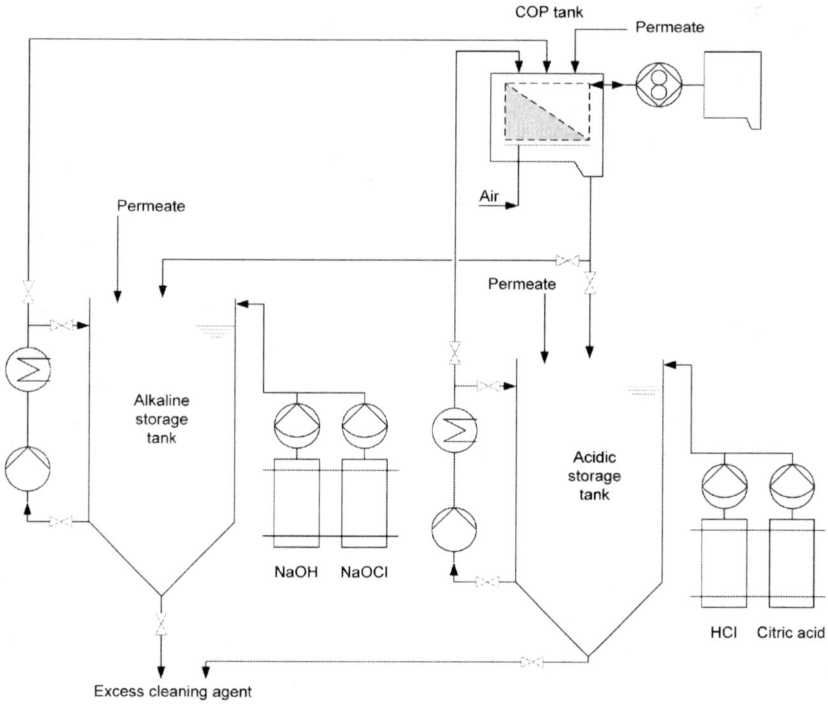

Figure 5.12 Simplified schematic diagram of the COP equipment

to undesirably high emissions of organic halogen compounds (AOX). Soaring concentrations of chlorine and the manual removal of braids and debris from the modules compromised the filters integrity. Thus, after less than two years of operation, the Erftverband decided to have the first line of original membranes replaced by a new generation less vulnerable to clogging. In 2003 the second filtration line was also replaced. The operation and layout of the Rödingen plant was thus modified through several years of research and development. In this process there was a systematic and pragmatic assessment of different cleaning concepts and chemicals at the plant, complemented by laboratory tests.

Cleaning Agents

In search of a substitute for hypochlorite, several commercially available cleaning agents were examined for their suitability to perform chemical cleaning at the Rödingen plant.

Hydrogen peroxide (H_2O_2) or a combination of hydrogen peroxide and peroxyacetic acid is used as a disinfectant in many industrial membrane applications. Under ideal conditions the residues from the cleaning are mainly water and acetic acid. Unfortunately H_2O_2 appears to be less effective in full scale application. The successful application of this cleaning agent has been reported at lab-scale conditions (Erftverband, 2004, Melin *et al.*, 2007) and at newly installed membrane filtration units (Wedi, 2006). Contrastingly, the full-scale experience at the Rödingen MBR revealed, that it did not yield the desired effect when used over a longer period and when applied on membranes that were already in service for several years (Erftverband, 2004). A possible explanation might be, that H_2O_2 is relatively unstable and shows a premature reaction in the permeate pipes and the membrane fibre lumen because of substances and the biofilm that can be found there. The deterioration of hydrogen peroxide can also be further catalyzed by metal-ions, that also could be found in the biofilm on the permeate side (Drensla, 2008).

A commercially available cleaner made of a combination of hydrogen peroxide and peroxyacetic acid only encouraged a modest increase in permeability at a high consumption rate during full-scale tests at the Rödingen MBR. Another argument against the application of this cleaner on a large scale was its unacceptable risk of explosion. Expensive safety precautions are required whenever this compound is stored in larger quantities.

Enzymatic cleaning agents required permanent and complete mixing of the bulk, but recirculation in the cleaning tank was not sufficient to reach all fibres within a cassette, while the use of membrane aeration triggered immense foaming. When aeration was stopped, the cleaning effect remained limited to the outer layers of fibres in the cassettes. Furthermore, high cost ruled out the enzymatic agents for large scale application.

Foulants

Fouling is still one of the major issues of the membrane bioreactor (MBR) research. Most attention is currently paid to extracellular polymeric substances (EPS) in either bound or soluble/colloidal (soluble microbial products, SMP) form (Drews *et al.*, 2008). While several trends or correlations were reported, the comparability of results is still limited by the numerous differences in plant set-up and analytical methods. Governing principles and the nature of inter-relationships between MBR parameters and fouling can be appreciated, nevertheless actual operating conditions and the associated absolute operating values can generally only be arrived at heuristically (Judd, 2006).

In the case of the Rödingen MBR an attempt was made to understand the nature of the foulants under the specific operating conditions in order to find bespoke cleaning chemicals and strategies (Erftverband 2004). Liquid chromatography-organic carbon detection (LC-OCD) was performed on the raw wastewater, the activated sludge, the permeate and extracts from the membranes. Polysaccharides and humic substances were identified as the major groups of foulants. In the laboratory most polysaccharides could be extracted at pH 7 while humic substances could only be extracted at pH 10, as can be seen from Figure 5.13. Unfortunately, the nature of the foulants was so diverse that, finally, this effort did not lead a way to find more specific cleaning agents.

Cleaning Protocols

A detailed study of different cleaning protocols was performed at a laboratory to screen for procedures with the highest potential for application. To determine the influence of various factors on cleaning efficiency, comparative tests on single fouled membrane fibres from the Rödingen MBR were carried out in a test apparatus. In these tests hydrogen peroxide or hypochlorite were used for oxidative cleaning and citric acid for the acidic cleaning step. The cleaning test protocols also varied in the sequence and duration of the cleaning steps performed in air or in water (but not in biomass). Cleaning the membrane samples in air almost doubled the permeability gain compared to a cleaning in water as long as equal concentrations of cleaning agents and equal durations for cleaning were used (Figure 5.14).

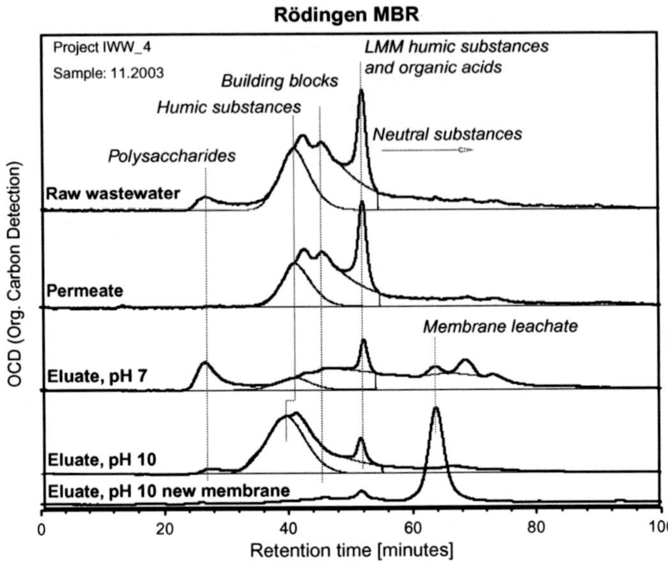

Figure 5.13 LC-OCD diagram of water samples and membrane eluates at Rödingen MBR

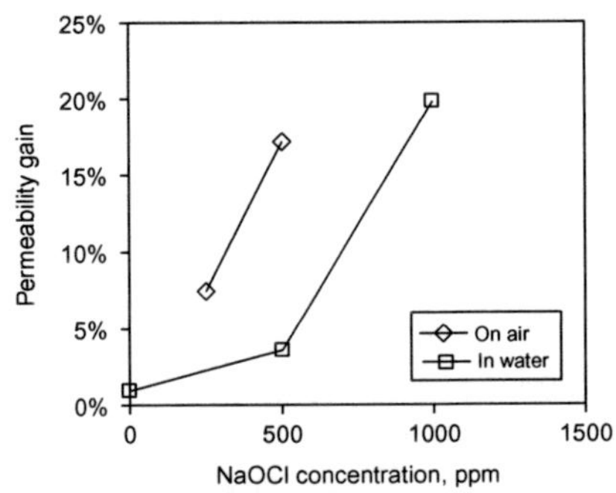

Figure 5.14 Comparison of lab-scale cleaning results (Erftverband, 2004) using CIP in water and in air at different NaOCl concentrations

Further on, at lab scale, there was no significant difference in outcome when the sequence of cleaning steps was changed, such that the acidic cleaning step was performed first instead of the oxidative step. On the other hand, as described below, the sequence of cleaning steps has a large influence in full scale installations. Table 5.4 gives an example of an optimized CIP cleaning protocol.

Table 5.4 Example of a weekly CIP cleaning protocol

Step	Description	Duration
1	Shut one filtration train	
2	De-aeration of filtration line	
3	Empty filtration tank (facultative)	1–4 hours
4	Interval-wise backwashing with chemicals	Approx. 2 hours
4.1	Acidic cleaning (Citric acid + HCl)	7 steps, backwashing 30 seconds, soaking 300 seconds
4.2	Permeate backwash	1,000 seconds
4.3	Alkaline cleaning (NaOCl + NaOH)	7 steps, backwashing 30 seconds, soaking 300 seconds
5	Refill tank with biomass (facultative)	1 – 6 hours
6	Start air scouring and biomass recycle	
7	Resume filtration	

LONG TERM EFFECTS

During the first months of operation of the Rödingen plant there was only regular enhanced backwashing with sodium hypochlorite at 200ppm. When it was discovered that there was also a massive coarse fouling and clogging of the membranes the Erftverband started using an external cleaning device for intensive COP (Brepols *et al.*, 2007).

Regular COP and Chemically Enhanced Backwashing

For more than 3 years the cleaning strategy relied mainly on two COPs per year, carried out before and after the winter season, and an intermediate chemically enhanced backwashing (CIP) using only sodium hypochloride. Figure 5.15 shows the saw-tooth variation of permeability which was typical for this cleaning strategy. The arrows indicate the intensive cleaning. The unusually high peak permeability observed at the first three COPs suggests that the cleaning also changed the membrane properties in such way as to probably compromise their function in the long term. Later this effect disappeared by limiting the amount of cleaning agents and monitoring the cleaning process more closely.

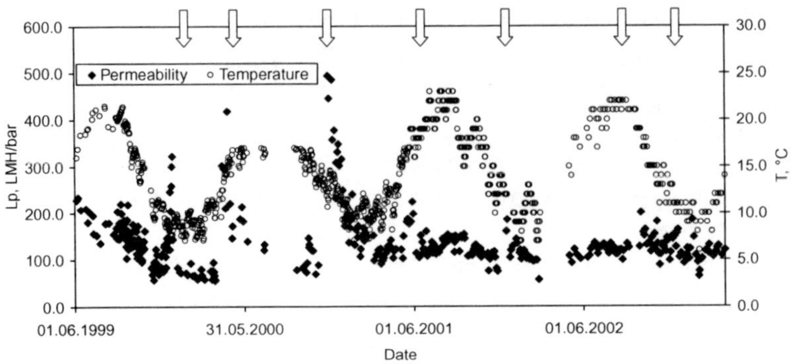

Figure 5.15 Filtration performance at Rödingen MBR, 1999–2003. The arrows indicate intensive cleaning by COP

Initially the intermediate CIP in biomass was performed with up to 1,000 ppm of hypochlorite every two to ten days. In 2001 a regular second step, employing citric acid, was added. This procedure maintained a higher membrane performance between the COPs.

COP at the Rödingen MBR is very labour intensive. The original plant design did not include COP equipment. Each single filter cassette has to be lifted from the filtration tank using a stationary crane and is then transported using a mobile crane to a COP container located outside of the machinery building where it is soaked and cleaned for up to 24 hours. The membranes have to be washed out carefully with clear water before they are reinserted. It thus takes over two weeks to have all 12 membrane cassettes cleaned. After COP, the used cleaning chemicals are stored in a disused tank and slowly dosed to the influent. During that time, the hydraulic capacity of the Rödingen MBR is reduced. In order to operate cleaned and uncleaned membrane filters in a line the flow at the cleaned filter cassettes is reduced by inserting baffles into their permeate adapter.

Cleaning in Place in Air

The method was first successfully employed by Stowa (2001) at a pilot plant, but initial large scale trials at Rödingen were not efficient. Modifications of the plant equipment were required before the regular CIP in air proved effective. After that, every one to two weeks the biomass was drained from the filtration tank before backwashing first with hypochlorite and second with citric acid. Figure 5.16 shows that this method has maintained a sufficiently high level of nearly stabilized permeability with seasonal variations. Only a precautionary COP was finally undertaken in November 2006.

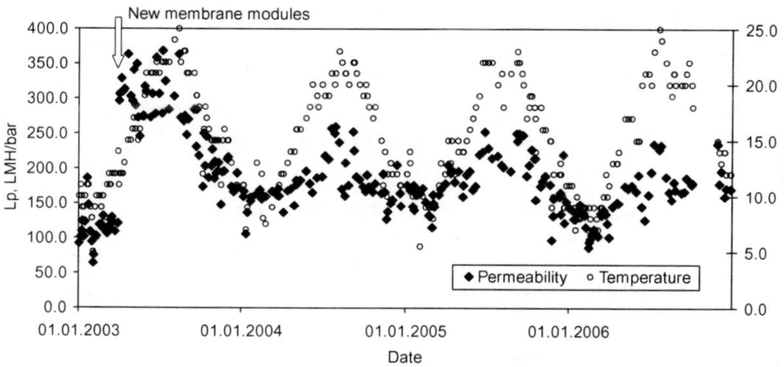

Figure 5.16 Filtration performance variations with CIP in air at Rödingen MBR, 2003–2006

The CIP in air cleans a whole filtration line in less than 6 hours, during which the most time is needed for draining and refilling the filtration tank. Additional equipment is needed for emptying the tanks. By removing the used cleaning agents from the tank, they can be treated separately.

CIP in Air, in Biomass or COP, Which One is Better?

Over the years several cleaning strategies have been tested. Table 5.5 summarizes some of their main characteristics. However it was found, that it was not always possible to transfer successful strategies from one plant to another.

At the Rödingen MBR two different strategies have been employed, each over several years: (a) COP twice a year with intermediate CIP in biomass; and (b) frequent CIP in air with an optimized cleaning protocol. In light of the goals of the optimization process, the CIP in air strategy was satisfactory, although it has to be admitted that previous changes in the plant layout also improved fouling control. The CIP in air reduced labour input and downtime during cleaning compared to COP. It maintained permeability at sufficient levels over a long period.

In view of the success at Rödingen, CIP in air was also installed at Nordkanal as normal procedure. The size of the filtration compartments at this much larger plant makes the draining and refilling more time consuming. CIP in biomass was first tried on two of the eight filter lines.

As the experience at Nordkanal shows, a similar cleaning effect can also be achieved with CIP in biomass (Figure 5.17) at the expense of more frequent cleaning and a higher chemicals consumption. There appears to be a trade-off between CIP in air and CIP in biomass in terms of longer plant downtime on one side and slightly higher cost on the other. Although, cost for chemical cleaning agents in any case was found to be less than 2% of operational expenses and thus is not the decisive factor. The potential of using these two methods in combination on a facility is not to be excluded.

Table 5.5 Main characteristics of the different cleaning methods

	CIP in air	CIP in biomass	COP
Additional equipment	Draining of filter tank, increased backwashing flux	No	External device, crane
Downtime of filter line	Relatively short 4–8 hours	Short 2 hours	Long >24 hours
Separate treatment of chemicals after cleaning	Possible	Not possible, chemicals cause biomass and effluent	Possible
Applicability	Only for some HF membrane types	For all backwash-able membranes	For removable membrane modules
Cleaning effect, permeability variations	Stabilized permeability when frequently used	Stabilized permeability when frequently used	High peak performance, saw-tooth variation

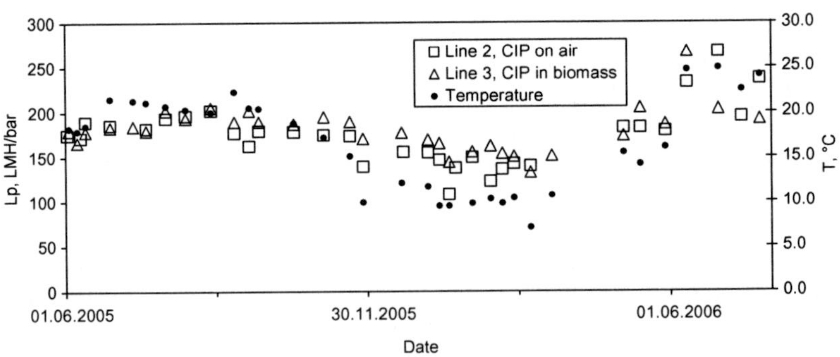

Figure 5.17 Filtration performance variations of Filter Line 2 with CIP in biomass and Line 3 with CIP in air, Nordkanal MBR

CIP in air works only with certain types of hollow fibre membranes. The lack of exterior water pressure during CIP in air constitutes an additional gradient in trans-membrane pressure along the membrane fibres during backwashing. This cannot always be tolerated as it may lead to incomplete backwashing and has to be compensated either by increasing the flux or lowering the water level gradually during the cleaning.

It was also found, that CIP in air was more effective at the Rödingen plant than at the much larger Nordkanal installation. The main difference in the cleaning procedure lies in the time that is necessary to empty the filtration basins. While emptying the tank, permeate that fills the lumen supposedly drains away by gravity, so that gas bubbles inside the fibres impede the following backwashing. This effect is aggravated by the time that is required for draining the filtration tank.

As a consequence, CIP in air seems to be more suitable for smaller installations with fewer filtration cassettes in a row and thus smaller filtration tanks that can be emptied more quickly.

Nevertheless, COP must still be carried out either as a precautionary measure at long intervals or as a remedial measure in cases of severe fouling or operational disruptions.

It can only be emphasized, that it is extremely helpful to have different options of membrane cleaning at hand in order to react adequately to changing boundary conditions and decreasing hydraulic performance of an MBR.

Sequence of the Cleaning Steps

Initially the Erftverband carried out a two-step chemical cleaning where an oxidative cleaning was followed by an acidic cleaning step. After several years it was first detected at the Rödingen MBR that the initial backwashing with sodium hypochlorite in an alkaline solution lead to a significant increase in the TMP during the backwash, which then frequently lead to an automatic precautionary shut-down of the filtration line. Finally, it became impossible to resume the cleaning. After changing the sequence of the cleaning steps to acidic cleaning first followed by alkaline cleaning, this effect disappeared.

Melin (2007) have found that, when using oxidative cleaning agents (NaOCl and H_2O_2) on fouled membranes, gas formation on the permeate side is a significant factor that might impede the proper distribution of cleaning agents.

Gas bubbles might then also partially block membrane fibres leading to the observed TMP increase. Practically it was found, that an initial acidic cleaning washes away ferrous residues allegedly coming from the phosphorous precipitation and thus seemingly reduces the gas-formation inside the fibre lumen.

Plant Availability During Cleaning

The plant availability is also affected by the applied cleaning procedure and indirectly can be transferred into costs that differ for the strategy chosen.

Applying any cleaning procedure requires some form of redundant filtration capacity, because the actual membrane area is reduced during the cleaning. One way to resolve this problem is to split up the membrane filtration into several lines and using a more conservative flux for standard operation that allows the remaining lines to work at higher flux rates for several hours or even days while one or more filter lines are cleaned. In this case, costs occur for additional equipment, construction and filtration surface.

The Nordkanal MBR for example can be operated with only six of eight filter lines at maximum inflow for a sufficient period. For smaller plants, like the Rödingen MBR that have only two filter lines, it can be more feasible to limit the hydraulic capacity of the plant while one filter line is cleaned and buffer the sewage or perform cleaning only during times of low inflow such as dry weather periods. However, this strategy bears the risk of having to postpone the cleaning for a certain time due to adverse operational or weather conditions.

Alternatively, cleaning can be carried out one by one, by cassette or by module. This reduces the filter area to a much smaller extent only, resulting in a more moderate increase of the flux rate of the remaining membranes. To avoid instantaneous fouling of the freshly cleaned membranes their flux has then to be throttled when reinserting them into the process until the line is cleaned entirely.

Scale Effects

In the long term, it became obvious that the membrane filters in the large installations behave heterogeneously. Hydraulic distribution inside the filtration tanks, the membrane modules and the piping is uneven and air inside the membrane fibres or the piping induces additional hydraulic losses. Incomplete dilution of cleaning chemicals may further obstruct proper CIP.

These effects add to the usual decrease of filter performance and are amplified with the size of the installation and may go undetected or underestimated in short term pilot trials or lab scale tests.

FORMATION OF UNWANTED BY-PRODUCTS

The use of NaOCl as a cleaning agent leads to the formation of chlorinated organic compounds, that can be measured as an increased AOX concentrations in the plant effluent. The quantity of AOX formed depends on the operating conditions of the cleaning. In Germany AOX effluent concentration are usually regulated and monitored by the governments water authorities. Table 5.6 gives the respective effluent threshold and the natural AOX concentration in the plant effluent uninfluenced by any in situ cleaning of the membrane filters. It can be seen that the regulated values leave only very little room for discharge of additional AOX during chemical cleaning.

Table 5.6 AOX effluent concentration, requirement and uninfluenced average value

MBR plant	Rödingen	Glessen	Nordkanal
Allowed maximum concentration, μg/L	50	50	100
Measured average concentration, μg/L	37	22	27

The chlorinated substances that are identified as AOX are formed as a product of the sodium hypochloride used for membrane cleaning and the organic substances in the bulk. During the in-situ cleaning AOX can be produced in the membrane tank and on the membrane surface. As biofilm also grows inside the permeate system and appears to accumulate over longer periods of operation, part of the AOX may also be formed on the permeate side during backwashing.

Measurements during the start-up phase of the Nordkanal MBR have shown, that immediately after the CIP, permeate that is retrieved from the cleaned filtration line may exceed the permitted AOX concentration significantly for a period of several filtration cycles. The cleaning procedure was thus optimized by the following means:

(1) only one filtration line is cleaned at a time,
(2) the permeate pipes and filters are backwashed thoroughly with uncontaminated permeate from other filtration lines and before the line resumes operation,
(3) membrane aeration and sludge recycling are operated to dissipate AOX that was formed close to the membranes so that sorption with the biomass is stimulated.

A critical accumulation of AOX in the biomass could not be monitored. At Nordkanal the AOX concentration in the dewatered surplus sludge lies within a range of 160 to 230mg/kg oven-dry mass. According to German legislation, the critical value for sludge used as a fertilizer for agriculture (Klärschlammverordnung, AbfKlärV, 15. April 1992, Bundesgesetzblatt I S. 912) is 500mg/kg oven-dry-mass.

MEMBRANE MATERIAL AND STRUCTURE

To get an insight into the structure, morphology and materials composition of the membrane filters scanning electron microscopy (SEM) and X-ray photoelectron spectroscopy (XPS) were performed on samples of the membrane material. These analyses were aimed at detecting changes in the membrane properties, that were a result of normal operation.

The membranes under investigation were asymmetric reinforced hollow fibres made of hydrophilic PVDF. The analyses that were performed on virgin membrane material were repeated in 2003 and 2008 using samples of the GE Zenon ZW 500 C membranes that had been in operation at the Rödingen MBR. Figure 5.18 shows cross-section details of new membrane fibres. The feed side is oriented to the right. The thickness of the membrane material is approximately 200μm. Following from the feed side to the permeate side, three different morphological areas can be identified on the pictures. The surface is made of fine pores and is approximately 5–10μm thick. Then follows a layer with large caverns in a more sponge-like material. More to the permeate side the porosity of the material increases slightly and the irregular imprint of the bearing fabric becomes visible.

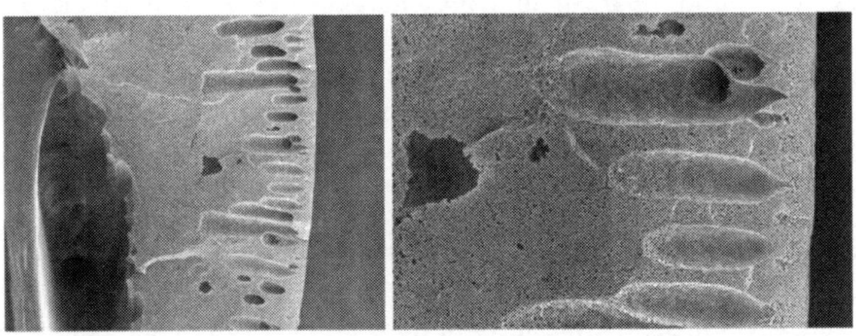

Figure 5.18 Full cross section of a virgin membrane fibre from the permeate to the feed side (left) and a detail of the feed side (right)

The comparison of the pictures of the virgin and used membrane (Figure 5.19) shows a slight compaction and deformation of the material on the feed side, while apparently no difference is visible between the membrane of different ages. It can be concluded, that the compaction of the surface layer occurs relatively quickly after commissioning as result of the filtration process. Figure 5.20 shows top views of the membrane surface at different stages. The visually observed surface roughness increases with life.

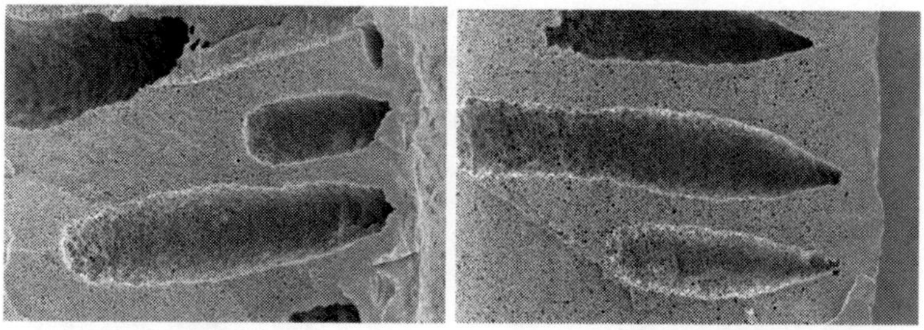

Figure 5.19 Cross section of a used membrane fibre after 2 years (left) and after 8 years (right) operation

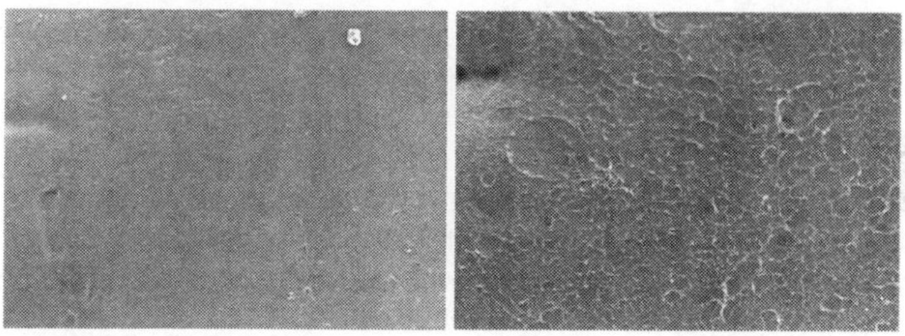

Figure 5.20 Top view of the membrane surface of a new (left) and a two years old membrane (right)

X-ray photoelectron spectroscopy (XPS), also known as electron spectroscopy for chemical analysis (ESCA), was used to determine the quantitative atomic composition and chemistry of the membrane material. It is a surface analysis technique that provides chemical state information of the detected elements. The process works by irradiating a sample with monochromatic x-rays, resulting in the emission of photoelectrons whose energies are characteristic of the elements within the sampling volume. When analysing organic polymeric membranes the energy spectrum of carbon is important to detect changes in membrane material composition. The full spectrum of carbon can be fixed by five single peaks C1 to C5. In the analysis, C1 is an equivalent to aliphatic hydrocarbons, usually coming from hydrophilic components of the membrane. C2 and C5 belong to the hydrophobic PVDF membrane material itself where C2 is an equivalent to the hydrocarbon components and C5 to fluorocarbons. C3 and C4 are of subordinate relevance (Erftverband, 2004).

The comparison in Figure 5.21 reveals, that the amount of hydrophilic components in the membrane surface decreases with time. A significant decrease of the percentage of one component automatically leads to an increase of the percentage of other components. In absolute values the amount of PVDF remains nearly constant.

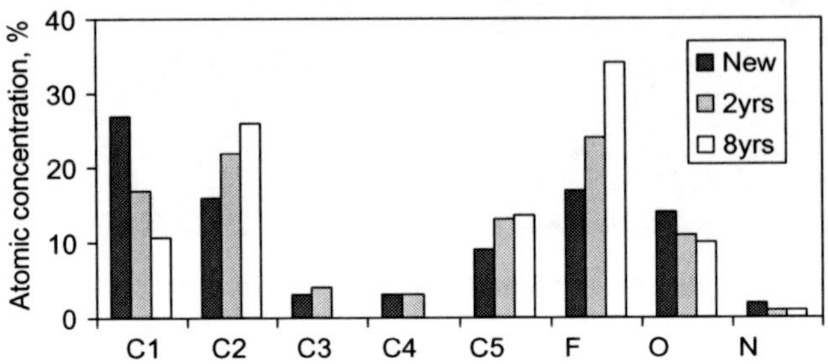

Figure 5.21 Atomic concentration of elements in the material of the membrane surface (feed side), based on Erftverband (2004)

Membrane Development

Membrane filter materials are subject of a permanent development by the membrane suppliers. The Erftverband has seen significant changes in the membrane morphology and structure over the years, even when the membrane filters came under the same type specification.

Originally the GE Zenon ZW 500 C membranes were strongly asymmetric with large caverns underneath the active surface (see above). ZW 500 C membrane filters that were delivered later on showed a more symmetrical and sponge-like structure (see Figure 5.22). The material seems to have a denser pore structure on the surface and should thus be less vulnerable to mechanical ruptures and the formation of pin-holes, while the missing caverns increase the resilience to compaction during operation.

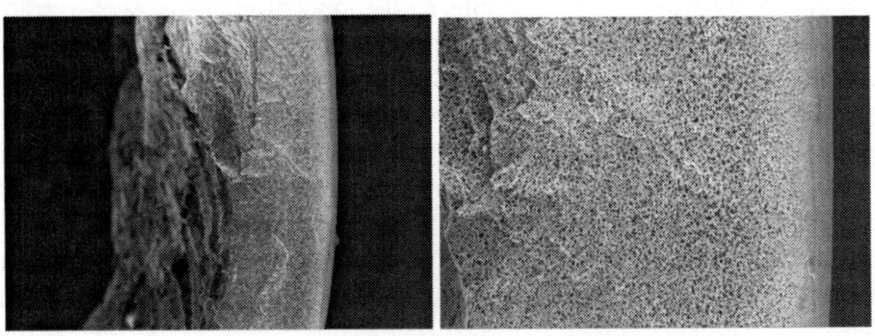

Figure 5.22 SEM photography of a new generation of GE Zenon ZW 500 C membranes (Erftverband, 2004), full cross section of a fibre segment (left) and detail of the feed side (right)

LIFE TIME

Once an MBR operation is stabilised after start up, it is possible to carry on with the same membrane modules for several years. It could be seen, that during their technical lifetime membrane filter materials are subject to changes in their chemical and mechanical properties. The same can be said for the properties of all adjacent parts of the membrane modules and cassettes such as headers, piping, fittings etc. Although these alterations influence membrane performance and possibly also the interactions between the membrane surface, foulants and cleaning chemicals, there are yet no threshold values that could help in predicting the end of service life.

De Wilde *et al.* (2007) have projected a service life of 13 years before filtration performance decreases to unsustainable levels, based on observation of the permeability decline and model calculations for the estimation of the impact of irreversible fouling. Anecdotal evidence from other MBR operators shows that the same membrane filtration units have been operated for 10 or more years.

Following scenarios could be used to determine at which point membrane replacement becomes inevitable:

(1) end of guarantee period,
(2) irrecoverable loss of permeability,
(3) irrecoverable loss of free membrane surface by clogging and braid formation,
(4) irrecoverable loss of other important properties such as effluent quality or TMP limits,
(5) integrity rate or number of leaking fibres,
(6) defects of adjacent parts of the membrane modules (e.g. aerators, headers),
(7) increased cleaning effort,
(8) increased maintenance and repair costs.

Proper membrane cleaning and maintenance help to postpone the end of service life. Even membrane damage can be remedied by sealing-off broken fibres, repairing broken joints or gaskets in the modules or even replacing whole modules within a cassette. All this has been carried out at the Erftverband's MBRs.

The example of the Rödingen MBR shows that the membrane replacement of one filtration line was triggered by problems with the chemical in-situ cleaning. The required concentrations of NaOCl in the regular weekly cleaning had to be increased to achieve sufficient cleaning results. As a consequence, the AOX concentration in the effluent exceeded the regulated values. Thus a reliable operation within the regulatory limits seemed no longer possible.

Finally a presumed trade-off between prolonged membrane life and increasing cleaning and repair cost would define a point when it becomes uneconomic to carry on with the operation of the same membrane filters. However, most processes that alter the membrane performance accelerate only slowly and gradually. The end of service life therefore has to be assessed individually based on technical and economical considerations.

REFERENCES

Brepols C., Drensla K., Janot A., Trimborn M. and Engelhardt N. (2008). Strategies for chemical cleaning in large scale membrane bioreactors. *Water Sci. Technol.*, **57**(3), 457–463.

De Wilde W., Thoeye C. and De Gueldre G. (2007). Membrane life expectancy assessment after 3 years of MBR operation at WWTP Schilde. *4th International Water Association Conference on Membranes for Water and Wastewater Treatment*, 15–17 May 2007, Harrogate, UK.

Drensla K. (2008). Verfahren und Erfahrungen mit der chemischen Membranreinigung, Membrantage, Gemeinschaftstagung von DVGW und DWA, Procedures and experiences in chemical membrane cleaning, 17.–19. Juni 2008, Kassel, Germany.

Drews A., Vocks M., Bracklow U., Iversen V. and Kraume M. Does fouling in MBRs depend on SMP? *Desalination*, **231**(1–3), 141–149.

Erftverband (2004). Optimierung einer Belebungsanlage mit Membranfiltration. Report to MUNLV NW, Erftverband, Bergheim, Germany, Optimisation of an activated sludge plant with membrane filtration.

Judd, S. (2006). *The MBR Book*. Elsevier, Oxford, UK, pp. 94–99.

Melin Th., Lyko S., Batsch A., Wintgens Th., Dott W., Al-Halbouni D., Hollender J., Pinnekamp J., Grinwis S., Schürmann B., Tacke D., Drensla K. and Janot A. (2007). Strategien zur Foulingkontrolle bei Membranbelebungsanlagen in der kommunalen Abwasserbehandlung, Abschlussberichtes für das Ministerium für Umwelt und Naturschutz, Landwirtschaft und Verbraucherschutz des Landes Nordrhein-Westfalen (MUNLV). Strategies on fouling control at membrane bioreactors for municipal wastewater treatment, Final report. http://www.lanuv.nrw.de/wasser/abwasser/forschung/pdf/ Abschlussberichtstrategien.pdf, Aachen 2007, accessed 1 July 2010.

Stowa (2001). *MBR for Municipal Wastewater Treatment, Pilot Plant Research Beverwijk WWTP*. Intermediate Report, Stowa, Utrecht, The Netherlands.

Wedi D. (2006). MBR Monheim, Operation with cold temperatures. *EUROMBRA Workshop – Biofouling in Membrane Systems*, 11–12 July 2006, Trondheim, Norway.

Wintgens Th., Rosen J., Melin Th., Brepols Ch., Drensla K. and Engelhardt N. (2003). Modelling of a membrane bioreactor system for municipal wastewater treatment, *J. Membr. Sci.*, **216**, Elsevier, 55–65.

Chapter 6

Investment costs

Christoph Brepols

INTRODUCTION

Building costs of WWTPs are influenced by a variety of factors. Some are related to the topography and geology of the building site, while others relate to the scale of the application, the technological equipment, aesthetic requirements or the market situation in the construction sector. As a consequence, it is often difficult to compare costs from site to site or even country to country. The Erftverband's 43 WWTPs are located in a region of approximately 1900km^2 (see Figure 6.1). More than two thirds of the WWTPs fulfil treatment requirements that exceed the European and German standard regulations. During the last 15 years many of the plants have been reconstructed or advanced treatment steps have been added.

COST STRUCTURE OF MBR PLANTS

The civil construction and the footprint of a MBR plant can be significantly smaller than for a conventional activated sludge plant with similar treatment performance. In the case of the Nordkanal MBR the footprint was reduced by more that 50% compared to an earlier planning as a conventional WWTP (Figure 6.2).

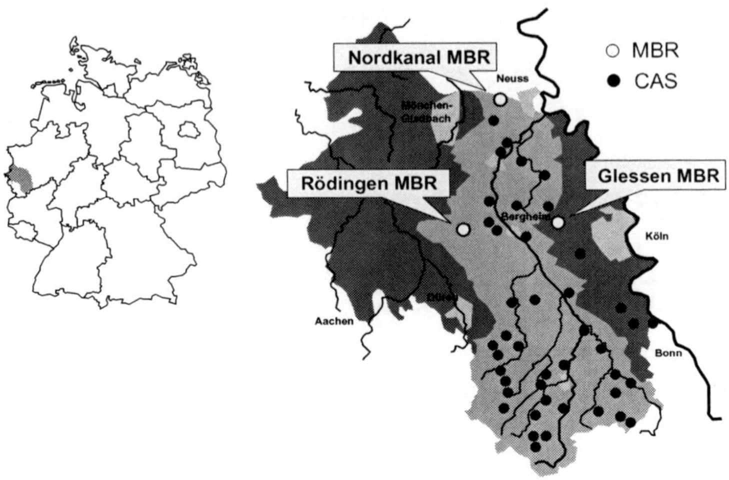

Figure 6.1 Location of the association area in Germany (left) and map of the Erftverband association area and location of the WWTPs (right)

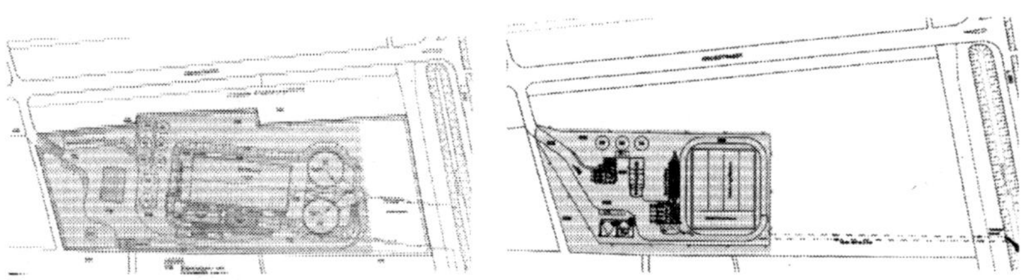

Figure 6.2 Site requirements of the Nordkanal WWTP, drafted as a CAS (left) and as an MBR (right)

At the same time the installation of the membrane filters requires additional equipment. As a result, there is a trade-off between constructional costs and the costs for mechanical and electrical equipment. While conventional plants have a greater share in constructional costs there is a shift towards the equipment that is mainly driven by the costs of the membranes and their peripheral equipment and the more sophisticated automation and process control of the membrane filtration. Wedi has published a share of 44% for mechanical equipment against only 39% for construction costs. These values resemble also the cost structure of the Erftverband's latest plants, Nordkanal MBR and Glessen MBR, as displayed in Table 6.1. The share of the membrane investment costs decreased significantly form 1999 (Rödingen MBR) to 2006 (Glessen MBR).

Table 6.1 Investment cost structure of three MBRs

Capital costs (MEUR)	Rödingen	Nordkanal	Glessen
Civil	0.6 (23%)	8 (37%)	2.6 (39%)
Mechanical	0.7 (27%)	6 (28%)	2.3 (34%)
Electric/automation	0.3 (12%)	1.5 (7%)	1.0 (15%)
Membrane system	1.0 (38%)	6.0 (28%)	0.8 (12%)
Total	2.6 (100%)	21.5 (100%)	6.7 (100%)

MEMBRANE REINVESTMENT

The membrane filtration units remain an expensive part of the equipment and their service life is limited. Reinvestment in the membrane filters thus plays a decisive role in the long-term economic success of an MBR. Based on an evaluation of the permeability values in a municipal facility, De Wilde *et al.* (2007) have projected a theoretical service life of 13 years. At the Rödingen MBR eight years of membrane lifetime were achieved under suboptimal conditions.

When the membrane filters were replaced at Rödingen in 2008, it was found that the limited compatibility between the membrane filtration systems of the different suppliers generated additional cost for changes in the peripheral equipment. This might also happen when using a new generation of membrane filters from the same supplier.

SPECIFIC INVESTMENT COSTS

Figure 6.3 shows the function of membrane investment versus the design flux and the specific membrane costs per square metre of membrane. The total (re)investment for the filters can be reduced either by increasing the design flux of the filters, which in fact reduces the required membrane area, or by lower purchasing costs. While the purchasing cost per square metre membrane area decreased from more than 200€/m^2 in 1999 (Rödingen MBR) to less than 70€/m^2, the design flux for the Erftverband's MBRs still does not exceed a value of 27 LMH for standard operation conditions at maximum inflow. The design flux is in the same order of magnitude for all three plants.

The specific capital costs in Table 6.2 illustrate the economy of scale as they decrease strongly with plant capacity. The capital cost values in Table 6.2 are based on the prospected values of the connection capacity and the average hydraulic load as the main design parameters for the plant layout. The specific membrane costs are calculated per square metre of membrane filters installed.

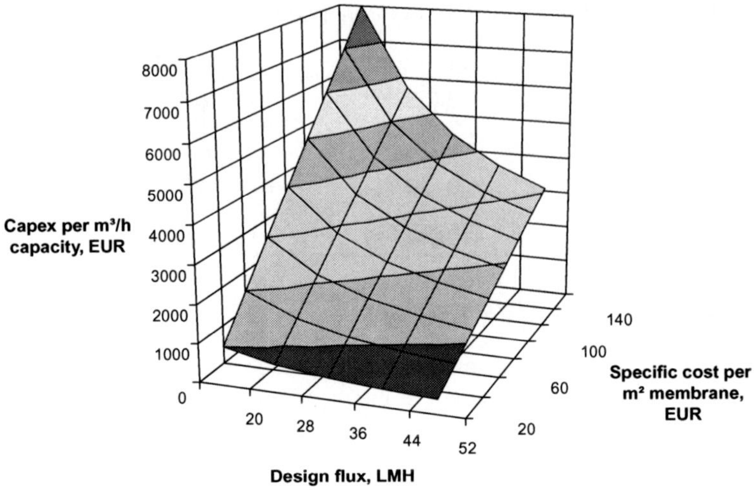

Figure 6.3 Cost function, specific membrane cost per m^3 treatment capacity

Table 6.2 Specific investment cost of the MBRs

Specific cost	Rödingen	Nordkanal	Glessen
Year of contract	1998	2000	2006
Capacity (PE)	3,000	80,000	9,000
Design hydraulic load (m^3/d)	450	16,000	2,400
Capital costs (EUR/PE)	867	269	744
Capital costs (EUR/m^3/d)	5,778	1,344	2,792
Membrane costs (EUR/m^2)	208	71	67

COST COMPARISON: MBR VS. CONVENTIONAL SYSTEMS

To facilitate the cost comparison between CAS, CAS with tertiary treatment and MBRs the investment cost data of 26 WWTPs from the years 1989 to 2008 were selected (Brepols *et al.*, 2010). During the period the construction cost indices for industrial buildings increased by 55.5% (Landesamt, 2008). The actual construction prices have been adjusted to correspond with these indices for the reference year of 2000.

The WWTPs were either built as greenfield installations or underwent a major reconstruction on site, resulting in a relevant increase of their treatment capacity or a significant improvement of their effluent quality. WWTPs that only underwent minor retrofitting are not included. The plants are all situated in the Erft river region and were built under the awarding authority of the Erftverband. The WWTPs share similar standards in civil construction, automation and mechanical and electrical equipment. The data include plants with simultaneous aerobic sludge stabilisation and plants with separate anaerobic sludge stabilisation. Plants with separate anaerobic sludge stabilisation are mainly found above a capacity range of 20,000m^3/d. The plants with simultaneous aerobic sludge stabilisation, including the MBRs, are designed for an SRT >25d. The tertiary treatment at CAS usually consists of a sand filtration and in some cases also a post-nitrification or post-denitrification. None of the CAS with tertiary treatment has to fulfil hygienic effluent regulations and there are no additional disinfection units in use. The data are shown in Figure 6.4.

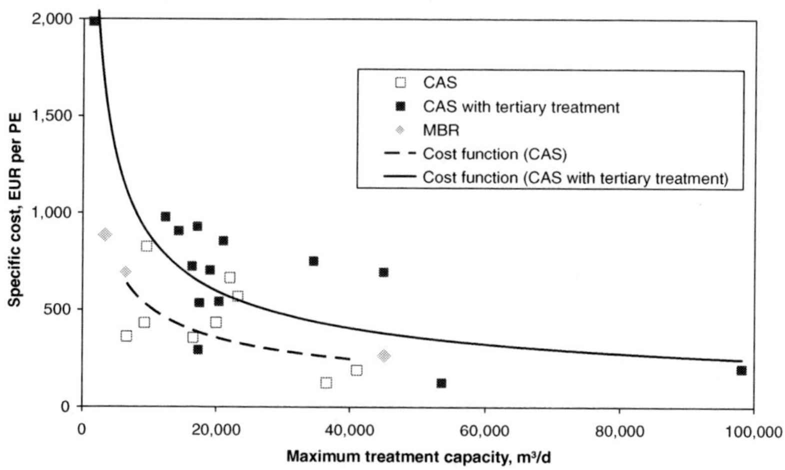

Figure 6.4 Cost functions for municipal WWTPs in the Erftverband association area

Two distinct cost functions were retrieved for CAS with and without tertiary treatment, each resembling a hyperbola. The specific costs of CAS with tertiary treatment are typically higher than the costs of other CAS. The MBRs are found in the range of the CAS without tertiary treatment. In terms of investment cost the MBRs thus show an advantage compared to CAS with tertiary treatment. It seems that the additional costs for the membrane filtration system are more than compensated by savings in civil construction costs.

REFERENCES

Brepols C., Schäfer H. and Engelhardt N. (2010). Considerations on the design and financial feasibility of full-scale membrane bioreactors for municipal applications. *Water Sci. Technol.*, **61**(10), 2461–2468.

Landesamt für Datenverarbeitung und Statistik Nordrhein Westfalen (2008). Ausgewählte Baupreisindizes in NRW (state office for data processing and statistics North Rhine–Westphalia, selected price indices for civil construction in North-Rhine Westphalia). http://www.lds.nrw.de/includes/php/druckversion.php?site=/statistik/datenangebot/daten/m/r323bauindex.html (accessed 5 August 2008).

De Wilde W., Thoeye C. and De Gueldre G. (2007). Membrane life expectancy assessment after 3 years of MBR operation at WWTP Schilde. *4th International Water Association Conference on Membranes for Water and Wastewater Treatment*, 15–17 May, Harrogate, UK.

Wedi D. (2003). Wirtschaftlichkeit des Membranbelebungsverfahrens (Economy of the membrane bioreactor process), ATV-DVWK Membrantage, 1./2., Juli, Bonn, Germany.

Chapter 7
Operational costs

Christoph Brepols and Heinrich Schäfer

INTRODUCTION

Operational costs at WWTPs are the result of different cost items. It can be assumed, that these categories apply universally. The actual amount of costs and the share they take in overall costs then depends on a variety of individual factors. These include the degree of plant automation, labour utilisation, the specific energy costs and waste disposal costs, as well as the company's cost accounting principles.

Figure 7.1 gives an example of the cost distribution at the Rödingen and Nordkanal MBRs. The Glessen MBR was excluded from this comparison because of its short run-time to date. The most significant cost items are sludge and waste disposal, personnel and energy costs. Operating fluids include precipitants, coagulants, lubricants and the cleaning chemicals for membrane cleaning. Maintenance costs also include repair of malfunctioning equipment and thus may vary significantly from year to year.

ENERGY CONSUMPTION

Energy consumption is a driving factor for the operational costs of membrane bioreactor plants. Figure 7.2 shows the actual average energy demand of the

Erftverband's WWTPs over a period from 2005 to 2007, including the two MBRs that were in operation at that time. The value for the Glessen MBR, which was started up in April 2008, was included based on data from the first year of operation.

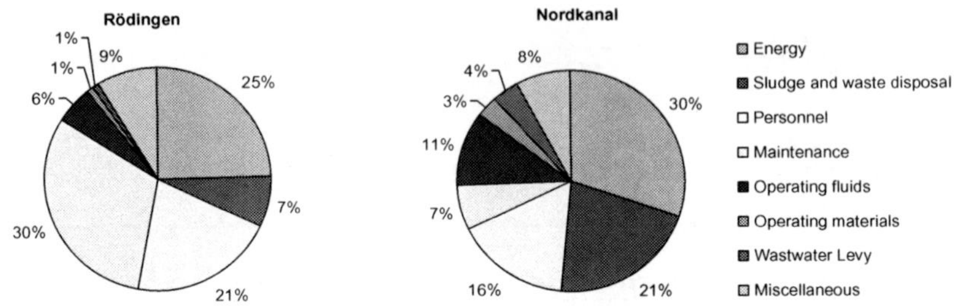

Figure 7.1 Share of the operational cost items (average of 2004–2008) of Rödingen and Nordkanal MBR

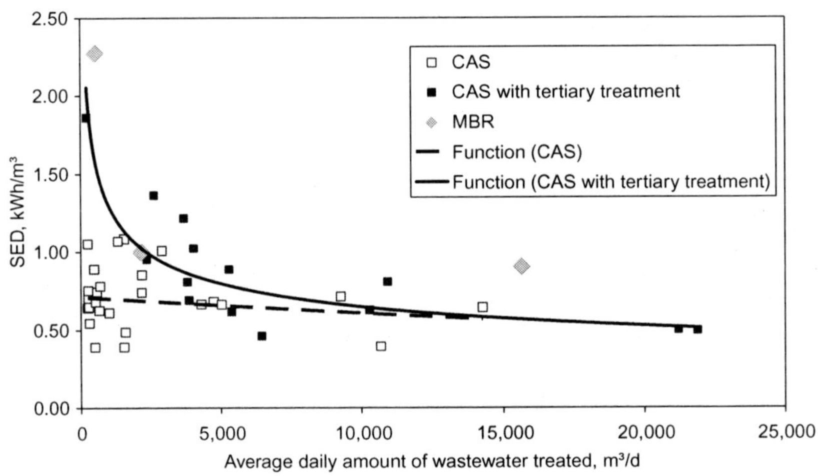

Figure 7.2 Actual specific energy demand (SED) of WWTPs in the Erft region

The energy consumption values in Figure 7.2 are calculated as the quotient of the total electric power uptake for the WWTPs, including all wastewater and sludge treatment and the actual amount of wastewater treated. At least for plants with a capacity of less than 5,000m^3/d there exists a significant difference in specific energy demand for CAS with and without tertiary treatment. As a consequence, two individual functions were derived. Furthermore, the data include energetically optimised and non-optimised plants.

Some plants actually receive dry weather loads that are significantly lower than their design value, leading to a disproportionately high specific energy consumption. These effects interfere with the usual economies of scale. The trend calculated for the CAS is less pronounced and seems to converge with the trend for CAS with tertiary treatment. Comparable guide values for energetically optimised CAS can be found in range from 65kWh a^{-1} PE^{-1} (<1000PE) to 27kWh a^{-1} PE^{-1} (>100,000PE) (MUNLV, 1999). This correlates to a range of approximately 0.9 to 0.3kWh/m^3 depending on the specific wastewater production per PE and the total annual amount of stormwater treated.

Figure 7.2 shows that the energy consumption of the MBRs with low or medium capacity is in the range of the CAS with tertiary treatment. For the large Nordkanal MBR the value in the graph is clearly above average. Because of its size, Nordkanal MBR competes here with CAS that employ anaerobic sludge treatment, have shorter SRTs and thus generally have a lower energy demand than WWTPs with simultaneous aerobic sludge stabilisation.

Pyöry (2009) conducted a field study to investigate the main energy consumers and the potential for energy savings at Nordkanal MBR. It turns out that the air scouring is responsible for 49.2% of the energy consumption. Process aeration in the bioreactor (12.1%), biomass circulation (1.3%), bioreactor mixing (11.5%) and permeate suction (2.7%) take significantly lower shares. The remainder (23.2%) is consumed by pumping stations, pretreatment, sludge dewatering and miscellaneous process units.

Figure 7.3 shows the energy consumption of the MBRs as a function of the daily inflow. Due to the mixed sewer systems in their catchment areas the plants have to deal with significant flow variations between minimum dry-weather and maximum storm weather conditions. These are more pronounced at the smaller two Rödingen and Glessen plants. When the membrane filtration is operated near its maximum design flux the specific energy demand decreases to 0.5kWh/m^3 at all three plants. The air scouring is operated at a constant air flow, regardless of the actual membrane flux. This is responsible for an increase in energy demand when the operating flux is low. One strategy that reduces energy consumption lies in adjusting the filtration area that is in operation according to the inflow by shutting down filtration lines. The Rödingen MBR is considerably underloaded during dry weather conditions and reflects an older level of knowledge in MBR technology, both factors explain the high specific consumption. The Glessen and Nordkanal MBRs have still potential to be optimised energetically. For example, Pyöry estimated for the Nordkanal MBR that a conventional optimisation of the given system by using energy efficient drives and reducing the runtime of certain machines would immediately reduce the specific energy demand by 15% while, under ideal conditions, an average energy consumption of 0.45kWh/m^3 could be achieved. Garces *et al.* (2007) and Tao *et al.* (2009) also report that a 35% to 40% reduction in energy consumption for air scouring can be achieved by applying a different aeration regime, while Tao et al. give a value of 0.4kWh/m^3 for an energetically optimised MBR with a capacity of 23,000m^3/d.

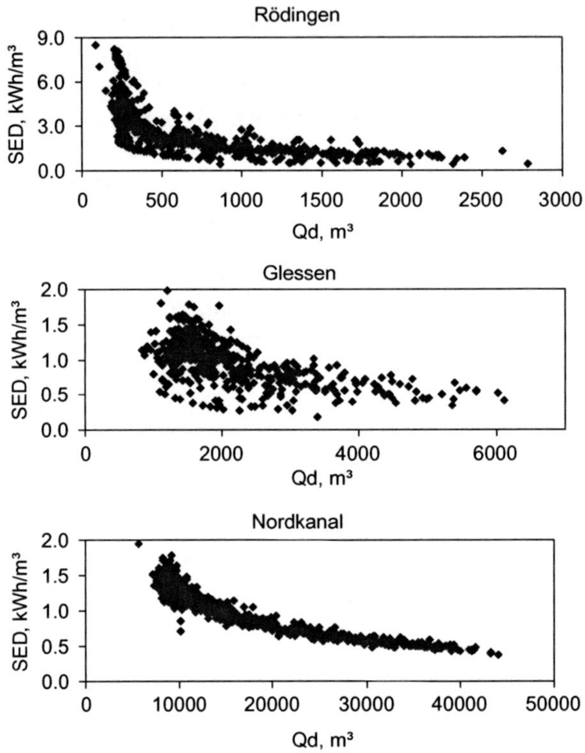

Figure 7.3 Specific energy demand and actual daily inflow

MEMBRANE CLEANING

At the three MBRs, weekly chemical maintenance cleaning in place (CIP) is performed in order to maintain stable permeability levels. Additionally, intensive cleaning out-of-place is performed once a year according to the manufacturers guidelines. The same chemicals are used for the different types of cleaning. Table 7.1 gives an overview of the chemicals used at Nordkanal MBR. The cleaning agents are diluted with permeate to reach their ready-to-use concentration. In-situ cleaning of the entire membrane surface of 84,400m^2 at Nordkanal MBR requires the amount of cleaning chemicals listed in Table 7.2 (Melin *et al.*, 2007).

Table 7.1 Specification of chemical cleaning agents raw solution

Cleaning agent	Concentration	Unit weight, kg/L	Costs, EUR/kg
Citric acid	50%	1.240	0.58
HCl	25%	1.125	0.23
NaOH	18%	1.219	0.28
NaOCl	12% (Cl^-)	1.220	0.24

Table 7.2 Cleaning chemicals consumption at Nordkanal MBR for a single CIP

Cleaning agent	Consumption, kg	Costs, EUR
Citric acid	992	575
HCl	248	57
NaOH	54	15
NaOCl	542	130

In total cleaning chemicals constitute a 0.5–2% share of the operating costs while an estimated 5–20% of annual personnel costs at the different plants can be related to membrane cleaning, most notably because of the time consuming and labour intensive COP cleaning (Brepols *et al.*, 2008). Optimised CIP generally allows the postponement of COP for longer periods.

SLUDGE AND WASTE HANDLING

In absolute values, the share of sludge and waste handling costs are comparable to conventional treatment plants as there are no significant differences in labour utilisation or the amount of waste produced. There is an increased amount of screening goods from the sieve facilities. On the other hand, these materials otherwise add to the excess sludge. At CAS with less intensive pretreatment these materials will simply end up unnoticed in the biomass and are then processed with biological surplus sludge. As the example of the Nordkanal MBR shows these fine sieve debris can be remixed with excess biomass before dewatering.

LABOUR

After commissioning, all three MBRs have been operated by the same staff as with their respective conventional predecessors. The smaller plants at Rödingen and Glessen are run as satellites of neighbouring larger WWTPs and are usually operated by a single worker, who is present during several hours a day to perform routine analyses and maintenance work. This is not different compared with similar sized CAS. At the smaller MBRs, the intensive ex-situ membrane cleaning regularly requires additional workers to handle the membrane filters. The Nordkanal plant is manned with six full-time workers, that are also responsible for the on-site sludge treatment and the connected pumping stations. Still the complexity of the process requires increased supervision.

All plants are only manned during a single daytime work shift. At night-time or during the week-ends, the plants usually run automatically. In the case of critical malfunctions an automatic on-site fault indication system alerts a service worker team.

COST COMPARISON: MBR VS. CONVENTIONAL SYSTEMS

In order to illustrate the differences in the cost structure and the influencing factors an example of a life-cycle-costs analysis is given. During preliminary planning it is common to perform cost comparison calculations in order to determine the most economic solution among various options.

A cost comparison method (LAWA, 2005) is employed to estimate the life-cycle costs of an MBR and a CAS and to investigate the influence of membrane service life on the economic efficiency of the MBR. It is assumed that the WWTPs have a capacity of 10,000PE, are greenfield installations, use biological nutrient removal and phosphorous precipitation and produce a comparable effluent quality. The MBR is designed with separate filtration tanks and hollow fibre membrane filtration. The CAS features sand-filtration and UV disinfection. A rough design is made for both installations assuming that the SRT is at 25 days and the MLSS concentration in the bioreactor is 4g/L at the CAS and 12g/L at the MBR. The calculated reactor volumes and plant footprints are displayed in Table 7.3. Property prices for acquiring a building site are not taken into account. The investment and operational costs shown in Tables 7.4 and 7.5 are estimated on the basis of empirical specific costs for the respective plant units that are typical for the Erft region.

Table 7.3 Key dimensions of the compared WWTPs

	Unit	CAS	MBR
Bioreactor	m^3	3 900	1 300
Secondary clarifier	m^3	800	–
Membrane surface	m^2	–	9 300
Sand filtration	m^2	30	
Paved surface	m^2	2 400	900

Table 7.4 Investment costs estimate

Investment costs	CAS, costs in EUR	MBR, costs in EUR
Civil works	4,960,097	2,248,507
Mechanical equipment	2,152,614	2,500,029
Electrical equipment	711,789	795,464
Membrane filtration	–	558,000
Total	7,824,500	6,102,000

Table 7.5 Annual operating cost estimates

Operating costs per year	CAS, costs in EUR	MBR, costs in EUR
Maintenance and repair	96,411	96,420
Waste and sludge disposal	25,664	25,664
Personnel	70,000	70,000
Energy	45,990	82,782
Operating fluids	11,000	12,860
Disinfection	7,665	–
Total	256,730	287,726

The net present values (NPV) of the MBR and the CAS are calculated for a 30 year period. The service life of the civil works at both installations is fixed at 30 years. For the mechanical and electrical equipment, excluding the membrane filters, the service life is 15 years. The cost comparison is performed by using different service life values of the membrane filters of 5, 7.5 and 10 years. It is assumed that the specific costs for membrane reinvestment remain constant over the years. Interest rates are fixed at 3%. The NPVs shown in Figure 7.4 are calculated based on the respective reinvestment cycles.

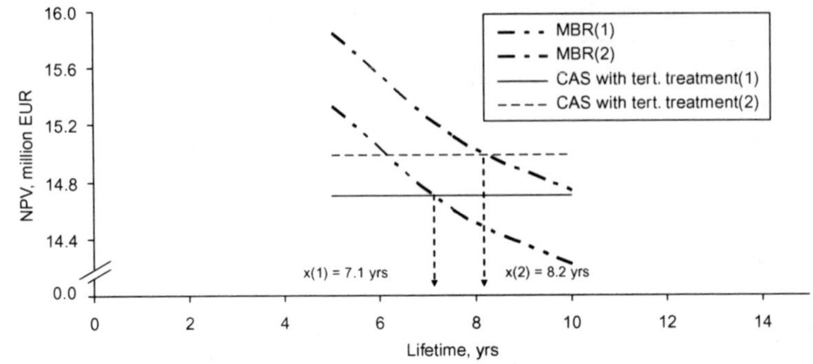

Figure 7.4 Net present values with constant energy costs (1) and increasing energy costs (2)

The increase in membrane filter lifetime from 5 to 10 years results in a EUR 1.1 million decrease of the NPV. It can be seen that the break-even point (x) of the MBR and the CAS under the given circumstances comes at a membrane service life of 7.1 years. A sensitivity analysis with energy costs increasing by 2% annually shows a shift of this point to 8.2 years. The example illustrates that under the specified conditions CAS and MBR can be equally economic when a reasonable membrane filter service life is achieved. This example should not be generalised. The precise economic characteristics of any decision between CAS and MBR need to be assessed individually. On a case to case basis, both monetary and also non-monetary aspects need to be taken into account.

REFERENCES

Brepols C., Drensla K., Janot A., Trimborn M. and Engelhardt N. (2008). Strategies for chemical cleaning in large scale membrane bioreactors. *Water Sci. Technol.*, **57**(3), 457–463.

Brepols C., Schäfer H. and Engelhardt N. (2010). Considerations on the design and financial feasibility of full-scale membrane bioreactors for municipal applications. *Water Sci. Technol.*, **61**(10), 2461–2468.

Garces A., De Wilde W., Thoeye C. and De Gueldre G. (2007). Operational cost optimisation of MBR Schilde. *4th International Water Association Conference on Membranes for Water and Wastewater Treatment*, 15–17 May, Harrogate, UK.

LAWA, Länderarbeitsgemeinschaft Wasser (2005). Leitlinien zur Durchfhrung dynamischer Kostenvergleichsrechnungen (KVR-Leitlinien) (Federal states working group water, Guidelines for the implementation of dynamic cost comparison calculations, edited by the LAWA subcommittee on economics in the water industry), Ausgearbeitet vom LAWA-Unterausschuss Wirtschaftlichkeitsfragen in der Wasserwirtschaft, /KVR-Leitlinien 2005, Kulturbuchverlag Berlin GmbH.

Melin Th., Lyko S., Batsch A., Wintgens Th., Dott W., Al-Halbouni D., Hollender J., Pinnekamp J., Grinwis S., Schürmann B., Tacke D., Drensla K. and Janot A. Strategien zur Foulingkontrolle bei Membranbelebungsanlagen in der kommunalen Abwasserbehandlung, Abschlussberichtes für das Ministerium für Umwelt und Naturschutz, Landwirtschaft und Verbraucherschutz des Landes Nordrhein-Westfalen (MUNLV) (Strategies on fouling control at membrane bioreactors for municipal wastewater treatment, Final report). http://www.lanuv.nrw.de/wasser/abwasser/forschung/pdf/ Abschlussberichtstrategien.pdf, Aachen 2007, accessed 1 July 2010.

MUNLV (1999). Handbuch "Energie in Kläranlagen". Herausgeber: Ministerium für Umwelt (Handbook 'Energy at wastewater treatment plants', Editor: *Ministry of Environment and Nature Conservation, Agriculture and Consumer Protection of North Rhine-Westphalia*) Raumordnung und Landwirtschaft des Landes NRW, Germany.

Pöyry (2009). Abschlussbericht zum Forschungsvorhaben Ermittlung energetischer Beurteilungskriterien für Membrankläranlagen mit Hohlfasermodulen und deren Verifizierung an einer Großanlage (Final report on the research project development of energetical assessment criteria for membrane bioreactor plants with hollow fibre modules and its verification on a large scale plant, awarding authority: Erftverband, Editor, Pöyry GKW GmbH). Auftraggeber: Erftverband Bergheim, Verfasser: Pöyry GKW GmbH, Essen, den 27.02.2009.

Tao G., Kekre K., Htun Oo, M., Viswanath B., Lew C. H., Kan L. M. and Seah H. (2009). Large scale membrane bioreactor plant design (retrofit) and optimisation. *5th IWA Specialised Membrane Technology Conference for Water and Wastewater Treatment*, 1–3 September, Beijing, P.R. China.

Chapter 8

Practical aspects of tendering and construction

Christoph Brepols and Heinrich Schäfer

TENDERING PROCESS

As a German company under public law, the Erftverband is obliged to operate within the public procurement regulations that enact EU directives governing the purchase of supplies, services and works by public sponsored organisations. They apply to the acquisition of works, supplies and services.

The aim is to:

- promote and enable free movement of goods and services between EU member states;
- maximise competition in contracts placed by public sponsored organisations;
- ensure open and accountable purchasing procedures and practices; and
- remove technical barriers to trade by harmonising technical standards across member states.

On the national level, the German construction contract standards (VOB) are thus mandatory for organising a tendering process for civil constructions such as wastewater treatment facilities.

In general the following award procedures can be and have been applied: (a) open procedures, (b) restricted procedures or (c) negotiated procedures with or without prior publication of a contract notice.

Example of a Tendering Process

The Glessen MBR, unlike the older two MBRs, was not built as a turn-key-installation. The tender was split up into several contracts:

(1) civil constructions (open procedure),
(2) mechanical pretreatment (restricted procedure),
(3) bioreactor equipment and membrane filtration units (restricted procedure) and
(4) electrical installations (open procedure).

To facilitate competition among membrane suppliers an attempt was made not to focus on a specific product beforehand. The tender documents therefore were partly described by functional datasheets in order to allow a variety of different membrane types to fit into the scheme. All other parts of the equipment were solicited on the basis of a deliverables list with unit pricing. Finally the tendering of the civil works took place after the contracts for the mechanical equipment had already been awarded to ensure that the size and layout of the filtration basins could be tailored to suit the actual membrane filters.

In contrast to this, the contracts for the whole installations at Rödingen and Nordkanal, including the civil works, were awarded as turn-key installations to general contractors. While the Nordkanal MBR was tendered in a restricted procedure, the contract for the first full-scale MBR in Rödingen was the result of a negotiated procedure after a pilot plant trial. The different procedures thus reflect the degree of knowledge within the Erftverband but also the situation and competition on the MBR market.

Turn-key contracts for MBRs can be advantageous for the awarding authority as they transfer the risk of improper interfaces within the works to the contractor but it also practically limits the number of appropriate bidders and the influence on many technical details.

On the other hand, the strategy chosen for the Glessen MBR required competence and experience to set up the detailed engineering. It also took away some liability from the membrane supplier. On the other hand the bidders in the membrane tender were able to focus more on their domain while at the same time an open procedure for the conventional parts of the WWTP achieved more competition. In the Glessen case this led to a significant reduction of building costs compared to the original cost estimate.

It is the organisation of the tendering process that leads to a trade-off between the degree of innovation and the risk of technical failure and influences the construction costs as well as the service-life of the project. Negatively this may lead to conservative designs, high prices and restrictions in plant operation while, positively, innovation, an open and clear assessment of financial and technological risks, lower prices and a higher potential for operational optimisation can be stimulated.

Some tools to manage this process from an awarding authorities perspective are:

- a plant design that leaves options for different membrane types or suppliers,
- clearly and openly defining the technical standards, quality and liability measures in the bidding documents and
- choosing the appropriate procurement procedure.

QUALITY CONTROL

In order to achieve a sufficient quality of the bid and the later work the tenderer needs to get a full picture of all circumstances that influence the design and operation of the membrane filtration.

Annex III to the EU Directive 93/37/EEC concerning public tendering procedures describes this as follows: 'In any case the contracting authority has to indicate, in the general or contractual documents relating to each contract, the technical specifications with which the works must comply. Technical specifications are the totality of the technical requirements contained in particular in the contract documents, defining the characteristics required of a work, material, product or supply, which permits a work, a material, a product or a supply to be described in a manner such that it fulfils the use for which it is intended by the contracting authority. These technical requirements shall include levels of quality, performance, safety or dimensions, including the requirements applicable to the material, the product or to the supply as regards quality assurance, terminology, symbols, testing and test methods, packaging, marking or labelling. They shall also include rules relating to design and costing, the test, inspection and acceptances for works and methods or techniques of construction and all other technical conditions which the contracting authority is in a position to lay down, under general or specific regulations, in relation to the finished works and to the materials or parts which they involve'.

In the case of the membrane bioreactor plants or membrane filtration units in particular the Erftverband provided a list of quality requirements with the contract documents that can be verified under actual operation (e.g. for acceptance tests) and that define criteria which may help avoid later disputes on the actual properties of the work.

The following standards and informations were provided with the contract documents:

- variations of the hydraulic load of the plant
- annual variations of the wastewater temperature

- information on the influent wastewater composition, conventional values like COD, nitrogen, pH-values but also any known substances that might be potentially harmful to the membrane material
- designed MLSS concentration in the bioreactor and possible variations
- effluent requirements including microbial pollution and AOX
- type and quality of the primary treatment
- information on operational standards and concepts and the required reserve capacity
- duration of the guarantee period

At submission, the bidders had to provide information on the properties of the membrane units with the tender:

- hydraulic capacity of the membrane, at start-up and at the end of the guarantee period
- materials, structure and morphology of the membrane
- pore size distribution
- mechanical, thermal and chemical resistance of the membrane material
- size, materials and details of construction of the modules and cassettes
- details on operation and cleaning

The acceptance test of the membrane filtration, that has to be carried out before commissioning, was already described with the tender. It included:

- optical inspection of the membranes
- bubble testing and pressure decay test of the membrane filters and the mounted filtration racks
- determination of the permeability under operating conditions
- test of the hygienic effluent quality

In order to assure a sustainable quality of the membrane filters the contract also stated that these tests are to be repeated at least once before the guarantee period ends.

GUARANTEE VALUES

Additionally, the quality of the membrane filtration was defined by several guarantee values.

- permeability at start-up and after a defined period,
- number of specified pathogens in the effluent,
- energy consumption,
- consumption of chemical cleaning agents.

Failure to comply with these values will result in penalties.

Chapter 9

Practical aspects of plant management and operation

Christoph Brepols and Kinga Drensla

OPERATIONAL AVAILABILITY

The operational availability of a membrane bioreactor is influenced by several factors. Some of these are also common to conventional activated sludge plants. Others, and these are the more critical ones, are linked to the properties of the specific process and the membrane filters. The membranes and the filtration process are vulnerable to a variety of external disruptions that come from the sewer system and internal disturbances and failures of design, equipment and operation of the plant.

Table 9.1 shows an overview of disruptions that have occurred during the operation of the three Erftverband MBRs, the secondary effects that were produced by the initial failures and the diagnostic and countermeasures that were taken.

EXTERNAL DISRUPTIONS

Harmful Substances from the Sewer System

The incoming sewage may contain substances that can be harmful to the membrane material or the biological process. They may come from industrial

Table 9.1 Overview of operational disruptions encountered

Trigger	Secondary effects	Diagnostic measures	Remedial measures
• Breakdown of peripheral equipment • Harmful influent • Improper membrane cleaning • Improper MLSS concentration • Insufficient membrane aeration • Insufficient pre-treatment • Leaking permeate piping • Low temperatures • Temperature drop • Membrane rupture • Module/cassette breakdown • Poor biomass mixing • Poor module hydraulics • Prolonged maximum hydraulic load	• Aerator clogging • Biomass impairment • Braid formation • Bulking biomass • Fibre ruptures • Foaming • Insufficient air scouring • Membrane deterioration • Module clogging • Solid deposits • TMP increase	• Bubble testing • Optical membrane inspection • Pressure decay testing • TMP/Permeability observation • Effluent quality measurement	• Improved pre-treatment • Increased airflow • Increased filter downtime • Increased recycle rate • Membrane cleaning • Membrane repair • Membrane replacement • MLSS adjustment • Module repair • Temporarily reduced inflow

discharges into the sewer system or accidents in the catchment area. The risk for membrane failure increases with the concentration, duration and amount of the inflow. The following groups of substances can be considered as potentially harmful:

- solvents, extracting or crystallizing agents
- thinners
- paints, varnishes
- adhesives
- pickling agents/solutions
- components for the production of plastics
- petroleum products (fuels, lubricants, oils)
- products containing silicone (e.g. cosmetic products)
- degreasing, cleaning agents
- disinfectants

In rural areas and at small plants the illicit disposal of liquid manure via the sewer system may cause operational disturbances by severely overloading the plants biological treatment capacity.

Effects. Water soluble substances primarily have an impact on the microorganisms in the plant and hence influence the performance of the biological treatment process. Non-water soluble substances may affect the membrane filters directly by wetting the membrane surface and (irreversibly) blocking the membrane pores, altering the membrane material chemically or damaging the membrane entirely.

Countermeasures. The Erftverband has taken different measures to reduce the risk of membrane failure due to harmful substances from the inflow. During the design phase an attempt was made to identify possible contributors of large quantities of such substances. It can be necessary to eliminate such contributions from the sewage by collecting and treating them separately (van Bentem *et al.*, 2007).

Municipal environmental authorities and local fire departments were briefed in order that the Erftverband is warned urgently if there is an accident with a spill of harmful substances in the catchment area of the plants. This would ensure that these substances are retained in the sewer before they reach the plant.

The Erftverband employs additional emergency buffer tanks on the MBR plants were harmful influents can be stored. Disused bioreactors or secondary clarifiers of the old treatment plants have been equipped to fulfil this task at relatively low cost. Instead of building new tanks, storage volume in the sewer or stormwater retention tanks could be used.

Other measures to prevent contact between any non soluble substance floating on the water surface of the membrane compartment and the membranes can be:

- installation and proper operation of a grease trap in the mechanical pre-treatment facility,
- installation of downflow baffles upstream of the membrane tanks
- increased water level above the membrane units

Temperature Drop

The Erftverband MBRs treat combined sewer effluent. Thus the wastewater temperature shows pronounced seasonal variations. In particular it has been observed that, during winter time, a sudden snow melt leads to a steep decrease of the influent temperature.

Effects. While permeability shows seasonal variations along with the temperature, a temperature drop leads to an abrupt decline of membrane permeability. The increase of the viscosity of water and changes in the biomass properties may also lead to increased deposits of biomass in the modules.

Countermeasures. A way to prevent such incidents is a preliminary intensive membrane cleaning before the winter season begins and a more frequent chemical backwash during the cold season. In advance it should also be checked that the membrane modules do not suffer from any clogging and the air scouring is working well.

Generally the cleaning strategy should aim at maintaining a maximum hydraulic capacity of the filters before temperature decreases. If that cannot be achieved an ad-hoc intensified ex-situ cleaning might become inevitable which then requires sufficient equipment to protect the membranes from cold weather and freezing.

Continuing High Hydraulic Load

The MBRs are designed for a maximum inflow during storm weather which is 2 to 3 times higher than the peak inflow with dry weather.

Effects. It can be observed, that long-lasting inflow at the maximum design load leads to a steady decrease of permeability and an increase in TMP. Usually the membranes recover from such events during times of low inflow. Short hydraulic peaks do not show any measureable effect while prolonged high inflow may even lead to a technical shut-down of the filtration because the TMP levels rise beyond their limiting values (e.g. TMP $>$ 0.2 bar).

Countermeasures. At the design stage, the risk can be reduced by choosing a more conservative design flow of the membranes filters and by following a multi-lane filtration concept. This allows intensified cleaning efforts on single filter lines while the plant remains in full operation.

Once in operation the following measures can be taken:

- increasing the aeration rate to improve air sparging of the modules,
- reducing the MLSS concentration in the membrane compartment either by reducing the overall MLSS content or increasing the recycle ratio between the (separate) filtration tanks and the bioreactor,
- more frequent chemically enhanced backwashing.

INTERNAL DISRUPTIONS

Braid Formation

The raw wastewater contains hair and fibrous materials. Once these materials get inside the membrane compartments of the plant they form braids on many parts of the equipment, the aerators and the membrane fibres (Figure 9.1).

Figure 9.1 Braid formation on the membrane fibres (left) and the aerators (right) as a result of insufficient pretreatement

Effects. These braids produce several negative effects

- mechanical damage of the membrane surface and rupture of membrane fibres
- obstruction of the air scouring of the membrane
- reduction of the fibre movement within the modules
- reduction of the free membrane surface
- accumulation of further debris in the modules

Countermeasures. It is crucial to remove the braid forming materials by means of a proper and adequate pretreatment. In the case of a failure of the pre-treatment braids can only be removed manually. As a first step, the membranes undergo an intensive chemical cleaning. This washes out biomass and grease from the braid and loosens their structure. Smaller braid can be washed out of the modules by spraying them with water. Larger braids have to be plucked out carefully after dismantling the membrane modules. It should never be attempted to remove braids by force as this will produce severe damage on the membrane surface. The biomass can be fed through the pretreatment in order to remove additional fibrous materials that are still in suspension.

Solids Deposition in the Membrane Modules

High MLSS concentration in the membrane compartment, low temperatures, frequent hydraulic influent peaks, poor hydraulics of the membrane modules and filtration tanks, inefficient or insufficient air scouring lead to solid deposits inside the membrane modules. It can be referred to as clogging, sludging or coarse fouling of membrane modules (see Figure 9.2).

Figure 9.2 Sludging of a membrane filtration module at Rödingen MBR, 1999, footer area of a filtration cassette (left) and detail (right). The dark color of the biomass inside the fibre-bundles indicates stagnant zones with no aeration.

Effects. The free membrane surface is reduced by the depositing of gel-like biomass that is mainly found on the membrane footer. In severe cases the biomass is dewatered, more solid-like and turns anaerobic. The permeability is reduced and the increased weight of the modules adds additional mechanical wear to the bearing construction.

Countermeasures.

- Reduction of the MLSS concentration in the filtration compartments
- Increased mixing in the membrane compartments e.g. by increased recirculation rates.
- Increased air scouring and improved maintenance of the aeration system
- Change of the chemical cleaning strategy
- More frequent chemically enhanced backwashing
- Intensive ex-situ membrane cleaning
- Regular optical inspection of membrane modules

Gas Bubbles and Particles in the Permeate

Filtration is achieved by pump suction with a suction pressure of up to 0.4 bar. Leakages in the piping, joints or the membrane cassettes, modules and fibres obstruct proper operation. Early detection of such failures helps prevent major breakdowns.

Effects. Through leakages below water level in the membrane tanks large particles get into the permeate system and when backwashing the membranes these particles are fed also to intact membranes leading to an increased blocking of pores or the membrane lumen.

Leakages above the water level bring additional air into the permeate system. The pumping efficiency decreases. The hydraulic capacity of the whole membrane filtration decreases. Pressure and flow in the permeate system are not measured correctly leading to flaws in the process control system.

Countermeasures. The permeate quality, the TMP and flow variations may be early indicators of any problems. If the values for a filter line start deviating from the plant's average this may hint at a bigger problem. In this case the line should be taken out of operation and investigated more closely.

Generally leakages can be detected by feeding pressurized air to the permeate pipes. The pressure has to be lower than the maximum allowable pressure by the membrane manufacturer in order to avoid additional damage.

Leakages below water level are indicated by air bubbles on the water surface (see Figure 9.3, left). Above water level suspicious areas of the equipment are sprayed with water or a mild detergent in order to see if air bubbles or foam appear.

Leakages are fixed, broken or leaking membrane fibres are cut out and sealed off according to the manufacturer's guidelines (see Figure 9.3, right). Permeate containing particles is removed from the system by purging it either with permeate from intact filtration lines or fresh water.

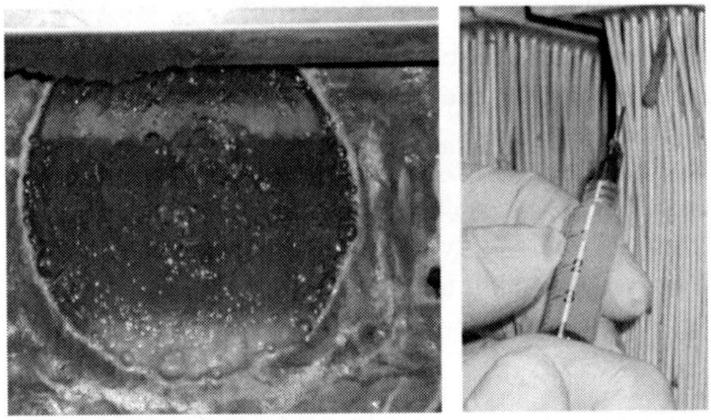

Figure 9.3 Bubbles indicating a leak in the module during integrity testing in the filtration tank (left) and sealing of a damaged fibre (right)

Improper Chemical Membrane Cleaning

Chemical membrane cleaning is a central remedial measure to cope with operational disruptions that affect the membrane performance, however, it can itself cause a variety of problems if not performed properly. The membrane material is able to tolerate a specified range of temperature, pH and chemical concentrations. The biomass also can be seriously harmed by too high concentrations of cleaning agents or residues from the cleaning process.

Effects. A carryover of larger quantities of cleaning chemicals (especially those with biocides) can contaminate the biomass and lead to a reduced biological treatment efficiency. At the Rödingen MBR severe foaming of the biomass was observed twice in the aftermath of intensified COP. Saponification of fat with sodium hydroxide was also observed, building up foam on top of the membrane filtration tanks.

High concentrations of cleaning agents (especially NaOCl) lead to an increased deterioration of membrane integrity. Currently it appears that an increased cleaning effect is achieved but in the long-term the membrane material ages much faster. An indicator of falsely applied cleaning can be permeability values that exceed the values observed with new membrane filters followed by a quick decline, so that the required cleaning intervals become shorter.

Countermeasures. Checking the concentrations and the temperature of the cleaning agents that actually reach the membrane surface is important. Incomplete dilution of chemicals, local temperature gradients or incomplete rinsing of residues from prior cleaning steps might cause conditions at least on parts of the membrane filters that differ significantly from the intended cleaning recipes. Unfortunately, these conditions can be very site-specific, depending on the equipment and size of the installation. Therefore, the only general recommendation is to check these procedures thoroughly regardless if they are carried out manually, semi-automatically or automatically on site and then recheck them again.

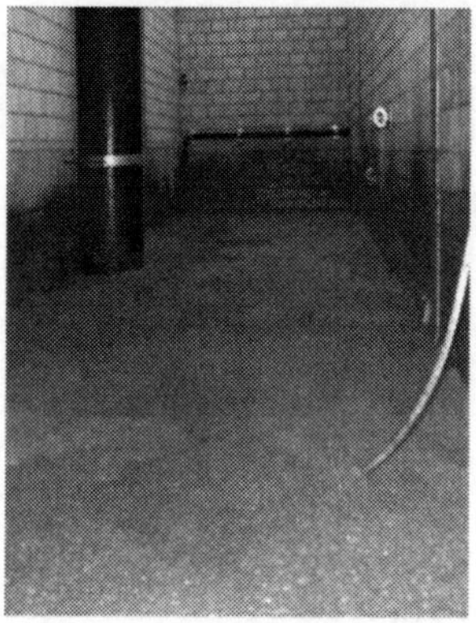

Figure 9.4 Foam building up on the filtration tanks after chemical cleaning at Rödingen MBR

FAILURES OF THE PERIPHERAL FILTRATION EQUIPMENT

The operational practice of the three MBRs has shown that the focus of operational disruptions is on the following MBR specific items. Process items, also common in CAS, such as pumping stations or the bioreactor aeration are not included.

Screening and Mechanical Pretreatment

As described in Chapter 3, the mechanical pretreatment is an intrinsic process of fail-safe operation of an MBR plant. In particular, the fine sieves, which are employed for the removal of fine hairs and fibres, can be a source of operational disruptions. Cleaning of the sieve surface and transport of the screenings can be critical. Overflow or bypass of these items in the case of operational disruption has to be avoided (see Figure 9.5).

Figure 9.5 Screen debris contaminating a membrane module after a breakdown of the influent screen

Membrane Aeration and Permeate Pumps

With the exception of the Rödingen MBR, which has only two filtration lines, the MBRs have been designed with extra filtration surface to allow continued operation in case of a breakdown of one or two filtration lines at higher flux. At low wastewater temperatures this operational mode is restricted to only a few hours or days at maximum inflow.

In order to reduce the cost of installation there is always only one blower and one permeate pump installed per filtration line. An increased maintenance effort is recommended.

Pneumatic Devices

Pneumatic devices are mostly used to drive valves and slides operating the membrane filtration. The service life of these units is defined by the number of switching cycles. Membrane aeration and filtration are discontinuous processes with often very short intervals. For example, the butterfly valves that facilitate the air cycling of the ZeeWeed membranes, switch every 10 seconds. A maximum number of one million switching cycles can thus be reached within six months and a breakdown of these devices leads to an immediate stop of a whole filtration line. Improper design and installation of the pneumatic drivers and the adjacent valves can add additional wear and tear. These devices thus require frequent maintenance. An on-site spares inventory to allow quick reaction in terms of a breakdown is also recommended.

Leakages

Leakages of the permeate systems above water level introduce air into the systems and allow the TMP increase. Below water level, biomass can become soaked and can contaminate the clean water side. Pipes and welded joints therefore have to be manufactured carefully.

Automatic Control System

The automatic process control of the membrane filtration is complex. It is designed to allow automatic operation of the whole plant during most of the time. Nevertheless, even careful programming and start-up cannot fully exclude occasional problems. Complex routines such as the chemical membrane cleaning are carried out under surveillance of an experienced operator only.

STAFF TRAINING

Fail-safe operation of an MBR requires a skilful and motivated work force. The complexity and sophistication of the new technological system can easily overstrain operators that are not accustomed to the process. Plant operation requires not only profound knowledge of the control procedures, it is also important to assess system performance correctly, identify the crucial factors and then take appropriate action.

Even with membrane suppliers offering occasional advice, start-up and operational optimisation of the first MBR at Rödingen required mainly learning-by-doing. Naturally this led to some more or less severe operational problems. In order to let others benefit from the lessons learned and to prevent them from repeating similar mistakes, the Erftverband very soon started to organise workshops on a national and international level to disseminate and discuss operational experiences among operators, consultants, regulators and researchers.

With an increasing number of MBRs being planned and going into operation an independent and non-profit training institute for membrane technology in wastewater treatment (SIMAS) was founded in 2005 with funding from the Environment Ministry of the German federal state of North Rhine Westphalia. Specialist and personal development of individual employees and companies involved in the design, construction and operation of membrane bioreactor systems for wastewater treatment is provided. The comprehensive theoretical and practical training in the operation of membrane bioreactor systems is meant to familiarize trainees with the use of different membrane systems. The courses concentrate on following practical aspects (Simas, 2010):

(1) principles of membrane filtration in wastewater treatment,
(2) mode of operation and differences between membrane bioreactor systems,

(3) terms and key figures,
(4) cleaning and maintenance,
(5) process monitoring and control,
(6) specific laboratory analyses for membrane systems.

MEMBRANE REPLACEMENT

Optical inspections of the filtration cassettes give evidence of any visible damages of the modules, aerators or frames (Figure 9.6). They can be performed either in an emptied tank or by removing single cassettes from the filtration tanks.

Figure 9.6 Inspection of membrane filtration units and bioreactor equipment in an emptied nitrification basin at Nordkanal MBR

Several membrane filtration cassettes at the Erftverband MBRs have been already repaired by replacing damaged joints, aerator pipes or membrane modules. Single membrane filtration modules can be replaced after cleaning the cassettes by means of a COP and then carefully dismantling them on site. This work is labour intensive can be carried out by experienced personnel only.

In 2008 a whole filtration line of ZeeWeed 500C cassettes at the Rödingen MBR was replaced with filtration cassettes of the KMS Puron type. Figure 9.7 shows the final assembly of the new filtration cassettes before inserting them into the filtration tanks.

Figure 9.7 Assembly of new filtration cassettes at Rödingen MBR

REFERENCES

Simas (2010). http://www.simas.de/ *Training institute in Seelscheid for membrane technology in sewage treatment* (accessed 29 March 2010).

Van Bentem A. G. N., Petri C. P., Schyns P. F. T and van der Roest H. F. (2007). *Membrane Bioreactors, Operation and Results of an MBR Wastewater Treatment Plant*. IWA Publishing, London, UK.

Chapter 10

Future prospects and required developments

Christoph Brepols

Optimisation of an MBR plant in practical operations is a permanent task with regards to cost efficiency, technical reliability and compliance with environmental standards. From an operators point of view, future developments in MBR technology would include the 'all-in wonder' membrane: a low cost, high flux, low energy, low fouling, self cleaning, durable, long-lasting filtration module. It is fully acknowledged, that at least some of the points on this wish-list are of a contradictory nature. However, compared with the situation more than 10 years ago, when the Erftverband decided to have its first full-scale MBR built, significant improvements have been achieved:

(1) improved module hydraulics and optimized process control have helped in reducing the energy consumption for the air scouring.
(2) advances in membrane production have led to improvements in the materials stability and resilience while market prices have come down.
(3) it could be seen, that in terms of investment costs, today's MBR plants are not necessarily more expensive than comparable CAS.

But there are also some drawbacks that have been encountered and problems that are not yet overcome.

(1) MBR plants are still very vulnerable to flaws in the mechanical wastewater pretreatment and this treatment step requires increased attention.
(2) The technology still has a huge potential for reducing the energy requirement. Further improvements in module hydraulics and a more case-sensitive and sophisticated process control could lead the way.
(3) Because of several technical shortcomings, membrane cleaning at full scale plants is often not as efficient as it could be. Gas entrapment in fibres and permeate pipes and unhomogeneous distribution of cleaning chemicals are the main concern at the Erftverband's MBRs.
(4) There is a lack of commonly approved procedures for the assessing of membrane performance and for the reliable detection of membrane failures, change of properties at full scale and critical values that could help to categorize the amount of damage and thus the need for action in membrane cleaning or membrane replacement.

The following demands can thus be summarized from a practitioner's perspective:

(1) fouling control and membrane cleaning to be further improved
(2) pretreatment for MBR plants should either become more reliable or be less important for the membrane filters
(3) energy consumption to be reduced
(4) membrane and module service life to be extended
(5) improved compatibility of membrane filters across product generations and suppliers to help limit the replacement costs.

Chapter 11
Summary

Christoph Brepols

During the last decade membrane bioreactor (MBR) technology has grown up to become state-of-the-art in municipal wastewater treatment. Since 1999, the Erftverband has designed, tendered and commissioned three full scale MBRs for municipal wastewater treatment in Germany, with capacities from 3,000 to 45,000m^3/d. Since then, operation and maintenance of the MBRs has become a daily function that is tied in with various drawbacks and successes.

The Erftverband was one of the pioneers in the full scale application of the technology in Europe and regularly hosted training and information workshops for plant designers and operators from all over the world, aiming at building confidence in the new technology. The Erftverband also collaborated in some major European research projects dedicated to MBR technology. This report provides hands-on information on many aspects of MBR technology based on more than ten years of practical experience in the design, commissioning and operation of MBR plants with submerged hollow-fibre microfiltration units.

Chapter 1 gives an overview on the hydrological, technological and policy background that motivates the employment of MBR technology.

Details on the process configuration and plant layout of the three MBRs are given in Chapter 2 together with some key recommendations on plant design.

Process data from all treatment steps of the three plants (mechanical pre-treatment, bioreactors, filtration, membrane cleaning) reveal an insight into the long-term performance of the MBR plants but also into possible do's and don'ts of full-scale applications. The efforts and potential of process optimisation are described. Chapter 3 is dedicated to the mechanical and biological wastewater treatment processes on the plants. The implications of MBR in nutrient removal and process control are described. Chapter 4 deals with the plant's practical influence on the receiving rivers. Chapter 5 then gives information on long term membrane filtration performance and chemical membrane cleaning. Membrane cleaning today mainly relies on a weekly maintenance cleaning in place using citric acid and sodium hypochloride in subsequent cleaning steps. The year-long quest for alternative cleaning agents and an optimised cleaning strategy is depicted.

Information on investment and operation costs based on the case studies but also in comparison with data from the conventional activated sludge (CAS) treatment processes complete the report in Chapters 6 and 7. It can be shown, that the MBRs have been competitive to conventional activated sludge plants with tertiary treatment in terms of investment costs. The different cost items, which define the plant's life-cycle cost are described with a focus on energy consumption. An example of a life-cycle cost calculation is given.

Some practical advice on tendering, construction, plant management and operation can be found in Chapters 8 and 9. A comprehensive list of the operational flaws that occurred at the Erftverband plants throughout the years can be used as a check-list for precautions in order to improve the safety of plant operations.

Finally Chapter 10 gives a short overview on some desirable developments in MBR technology from an operator's perspective.

Index

Note: Page numbers followed by *f* and *t* refer to figures and tables, respectively.

Lightning Source UK Ltd.
Milton Keynes UK

173859UK00001B/2/P

9 781843 393054